NORTH

EAST

SOVTH

The English o

The Spanishe Fleete

SEMPER EADEM

The Scale of English miles · Robert Adams authore

MAPPING
NAVAL WARFARE

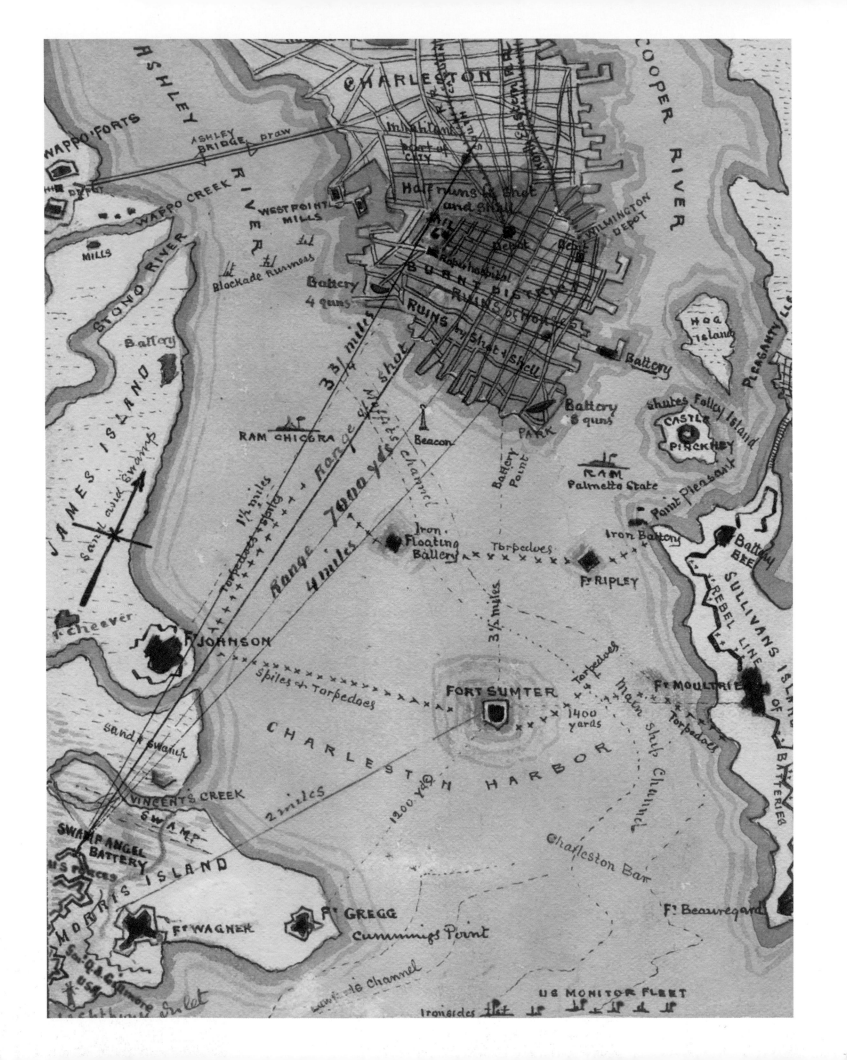

JEREMY BLACK

MAPPING
NAVAL WARFARE

A VISUAL HISTORY OF CONFLICT AT SEA

OSPREY
PUBLISHING

OSPREY

An imprint of Bloomsbury Publishing Plc

Osprey Publishing   Bloomsbury Publishing Inc.
Bloomsbury Publishing Plc 1385 Broadway, 5th Floor,
PO Box 883, Oxford,   New York,
OX1 9PL, UK   NY 10018, USA

www.ospreypublishing.com

OSPREY is a trademark of Osprey Publishing Ltd,
a division of Bloomsbury Publishing Plc.

First published in Great Britain in 2017

A CIP catalogue record for this book is available from the British Library.

Jeremy Black has asserted his/her right under the Copyright, Designs
and Patents Act, 1988, to be identified as the author of this Work.

ISBN: HB: 978-1-4728-2786-9
        ePub: 978-1-4728-3137-8
        ePDF: 978-1-4728-3138-5
        XML: 978-1-4728-2787-6

17 18 19 20 21   10 9 8 7 6 5 4 3 2 1

Index by Marie Lorimer

Designed by Nicola Liddiard, Nimbus Design

Printed in India through Replika

Osprey Publishing supports the Woodland Trust, the UK's leading
woodland conservation charity. Between 2014 and 2018 our donations
are being spent on their Centenary Woods project in the UK.

To find out more about our authors and books visit *www.ospreypublishing.
com*. Here you will find extracts, author interviews, details of forthcoming
events and the option to sign up for our newsletter.

Preface

Maps, war and the sea, this book brings together major interests of mine. Both text and maps combine to produce a fascinating account of naval warfare. The general emphasis in discussing naval warfare is on weapon systems and technological change, with change presented as a key enabler of capability. That is particularly important when writers look back from the period of rapid change over the last two centuries, rapid change that made other systems be, or at least appear to be, obsolescent. That account has value for the last two centuries, although perception played a major role in the understanding of effectiveness. Moreover, older ship types could retain value.

More serious, however, is the use of this approach to consider earlier periods and the extent to which, throughout, an emphasis on technology can lead to a serious underrating of other factors, notably organisational, particularly logistics, but also fundamental factors of command, fighting quality and tactics. Motivation, the maritime skill background and the long-lasting 'naval tradition' of a society are also important. The way a society looks at the army and navy may not simply be a matter of the length of its coast or other materialist issues.

A focus on technology and the habit of viewing the past in terms of transformative change have led to a search for military revolutions, and notably so to prefigure those discerned over the last two centuries. In particular, much stress has been devoted to a so-called 'military revolution' in the early-modern period, one indeed presented as lasting from 1500 to 1800. This, however, is not overly helpful. First, the idea of such a long period being described as a revolution is questionable, and certainly for modern history. Secondly, many of the key changes with reference to gunpowder in fact occurred prior to 1500.

Thirdly, for a long time gunpowder weaponry was adapted to existing weapon systems, rather than used to transform them. It provided different and additional firepower but could only work effectively at more than very short range in limited circumstances. The boarding of ships, a practice going back to the start of naval warfare, remained more significant than any emphasis on the 'ship killing' capacity of artillery might suggest. There has been a tendency to read back from conflict over the last century and a half to the earlier situation. Iron/steel ships sink more easily than their earlier wooden counterparts. Even when the artillery was very active and increasingly effective in the time of Horatio Nelson, the British still captured many ships of the line from the French and Spaniards at Trafalgar (1805). Artillery nevertheless could be linked to boarding as a sailing ship that had lost its rigging due to opposing cannon fire was in effect helpless, and thus vulnerable to boarding. Rigging was much easier to damage than the engine in the armoured belly of an iron ship.

The understanding of the use of naval power can also be overly simple. There is a temptation to think only of the gains to be made through the use of this power, namely acquiring resources through violence, and of trade as a power relationship. This tendency is valid, but it should be related to the mutually beneficial nature of trade, as well as to the need for maritime powers to accommodate themselves to the demands and opportunities of land-based economic systems, if they were to be much more than transient plunderers.

The category of transient plunderers, however, was important to naval power in much of the world. The use of ships for plundering, piracy and associated activities such as the seizure of people to sell as slaves remained highly significant. To a considerable extent, this use was stopped by imperial navies in the nineteenth century, notably by Britain's dominant Royal Navy, as they sought to guarantee the security of traders, on which the freedom of trade rested, but in parts of the world this instability has revived in recent decades.

The tendency to oversimplify also applies to naval mapping. A stress on change is valuable, but it needs to sit alongside an awareness of crucial continuities and key limitations. They, however, do not prevent what did occur from being fascinating.

It is a great pleasure to dedicate this book to Heiko Werner Henning, a most perceptive commentator on naval warfare, with thanks for repeatedly giving excellent advice.

Acknowledgements

For Heiko Werner Henning

This book brings together two of my central interests, war and cartography. Naval warfare has been a key element in global conflict, and has presented a major topic for mapping. Assessing both this mapping and naval warfare as a whole has been great fun to research and write. I am most grateful to my editor, Lisa Thomas, a great companion in the ways of illustrated books. While thinking about and working on this book, I benefited from the chances to sail the Atlantic, Baltic, Black, Caribbean, Indian and Pacific oceans or seas. I have benefited from the advice of Stuart Long, Wayne Malbon, Albert Nofi and Heiko Werner Henning on earlier drafts. Advice on particular points from Kenneth Swope, Harold Tanner and Arthur der Weduwen has been very useful. The opportunity to teach naval history at Exeter has also been most welcome.

CONTENTS

MEDIEVAL NAVAL WARFARE, FROM *LE LIVRE DES ANSIENES ESTOIRES*, C.1285 The crisis of Byzantine power after Constantinople fell to the Fourth Crusade in 1204 was followed by the creation of a series of competing Latin states and by the expansion of the Venetian empire. This illustration shows a clash between the Duchy of Athens, one of those states, and Venetians based in Crete. Trade financed naval activity. It was common to board rival warships, and the capture of prizes by that means was an important goal. Boarding involved close contact and there were many similarities as a result between conflict on land and at sea. Most naval conflict took place in inshore waters, because of the limited cruising range of ships as they had to stop to take on water and food, and because of the significance of seizing and protecting bases. Crete was held by Venice until conquered by the Ottoman Turks between 1644 and 1669 in a long war in which the Venetians failed to sustain a blockade of the Dardanelles. OPPOSITE

Long-range voyages were frequent for much of history, but without the voyagers having reliable or useful maps to guide them or record their journeys and without us having maps as evidence for them. It is difficult for societies relying on verbal communications to create 'written maps'. Two examples are the Polynesians of the southern Pacific and the Vikings of northern Europe. In both, although evidence is limited, it is clear that orientation was established in terms of wind direction, tides and the moon. Sailors who were not literate were able to create mental maps. The Polynesians, whose double-hulled canoes could tack into the wind and travel with relative safety across the open seas of the dangerous Pacific, also probably navigated by using stars and by observing the direction of prevailing winds and the flight patterns of homing birds. When sailing offshore, Polynesian sailors could read the changes in swell patterns caused by islands such as the Marshalls, and thereby fix their position.

Recent attempts to re-enact Polynesian voyages, many across hundreds of miles of open ocean, suggest that navigational errors that resulted from the islanders' methods may have cancelled each other out so that the navigator's sense of where his craft was may have been reasonably accurate. Furthermore, their use of the star compass to establish a position of dead reckoning meant that they did not need to concern themselves unduly with matters of distance. They regularly found islands in the Pacific without the help of detailed maps and modern navigation equipment.

That information could be recorded, which is a crucial feature in accumulating knowledge for maritime activity. Pilotage information was represented in charts of sticks (notably the midribs of coconut fronds) and of shells, which were studied by mariners before undertaking their journeys. The results were impressive, with the Polynesians sailing from New Guinea as far as Easter Island (*c.* 300 CE), Hawai'i and Tahiti (*c.* 400 CE) and New Zealand (*c.* 700 CE).

Although highly impressive, these voyages were, however, limited by the standards of what the West achieved from the sixteenth century. As a result, Western explorers were able to make genuine discoveries about the Pacific, especially the northern Pacific. The Polynesians understood their world, but its scope was not that of the oceans of the earth. Moreover, Polynesian navigational techniques and shipping were not suited to the temperate zone, to cold climates, or to carrying much cargo. This remained the case with the Polynesian voyagers encountered by European explorers in the eighteenth century. As the Polynesian islands were mostly self-sufficient, there was no need to carry bulky food, and thus to develop larger cargo-carrying ships.

In the sole Western transoceanic episode prior to the late fifteenth century, the Vikings sailed to North America via Iceland and Greenland in about 1000 CE. Their navigational methods are unclear. It has been suggested that crystal sunstones, which establish the sun's position as they enable the detection of polarisation (the properties displayed by rays of light depending on their direction), were used to navigate ships through foggy and cloudy conditions. However, their effectiveness has been contested. Viking longboats, with their sails, stepped masts, true keels and steering rudders, were effective ocean-going ships capable of taking to the Atlantic; yet they were also able, thanks to their shallow draught, to be rowed in coastal waters and up rivers, even in only three feet of water. They proved most effective in mounting amphibious operations and their crew, once landed, acted as troops. In modern terms, these ships were the first effective 'littoral combat ships'. As such, the

len par le comandement le roi
q cuidoit q ce fust ses fiz : plus
deuint cruel. Tsi qil manioit ho
mes & femes qit il les pooit ata
indre . Por cel deables enferer
manda le roi minos dedalus . T
si li fist faire vne maison mer
ueillouse . atant dentrees & de
chambres : q el mont nauoit
criature : se la dedenz en lamoie
ne fust encloz : q iamais fust re
paire alentree . Car . c . huis ia
uoit puis q len passoit le pre
meral : q tos desueoient al qui la
dedenz estoient . T encelle maiso'
fu al mostres encloz .

Coment al datheneo estoient
subget a cil de crete.

En cel tens estoient al dathe
nes si subget au roi minos
de crete : qil li deuoient enuoier chas
cun an . vii . uasles & vii . damoi
selles de treuage . Ttels come le
roi minos les mandoit ou roiau
me . Et qit il estoient uenus en
crete : le roi les faisoit metre de
uant son mostre qil auoit encloz
q les deuoroit sans atendance .
Adonqs en cel tens estoit egeus
roi dathenes : si li couint enuoier
son fiz theseus en crete por cel tru
age auuec les autres . Et quant

ZHENG HE'S ROUTE FROM NANJING VIA SOUTHEAST ASIA TO THE PERSIAN GULF, FROM *WU BEI ZHI* PUBLISHED C.1621 A stele erected by Zheng He (1371–1433) explained his mission as one of spreading the Chinese world order of peace that held barbarianism at bay: 'Upon arriving at foreign countries, capture those barbarian kings who resist civilisation and are disrespectful, and exterminate those bandit soldiers that indulge in violence and plunder. The ocean route will be safe thanks to this.' The ships were probably the largest wooden ships built up to then and, thanks to watertight bulkheads and several layers of external planking, they were seaworthy, but claims that they were nearly 400 feet in length have been questioned, not least as the dimensions do not correspond to the figures for carrying capacity, tonnage and displacement. During the seven expeditions sent into the Indian Ocean between 1405 and 1433, Zheng He reached Yemen and Somalia and successfully invaded Sri Lanka in about 1411, but claims that the Chinese circumnavigated the world lack credibility. RIGHT

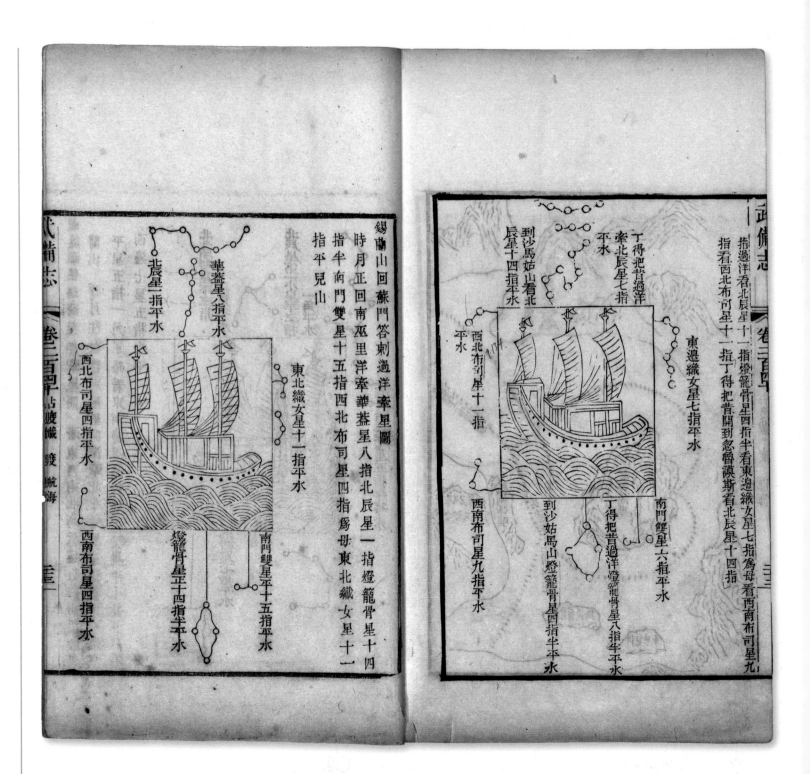

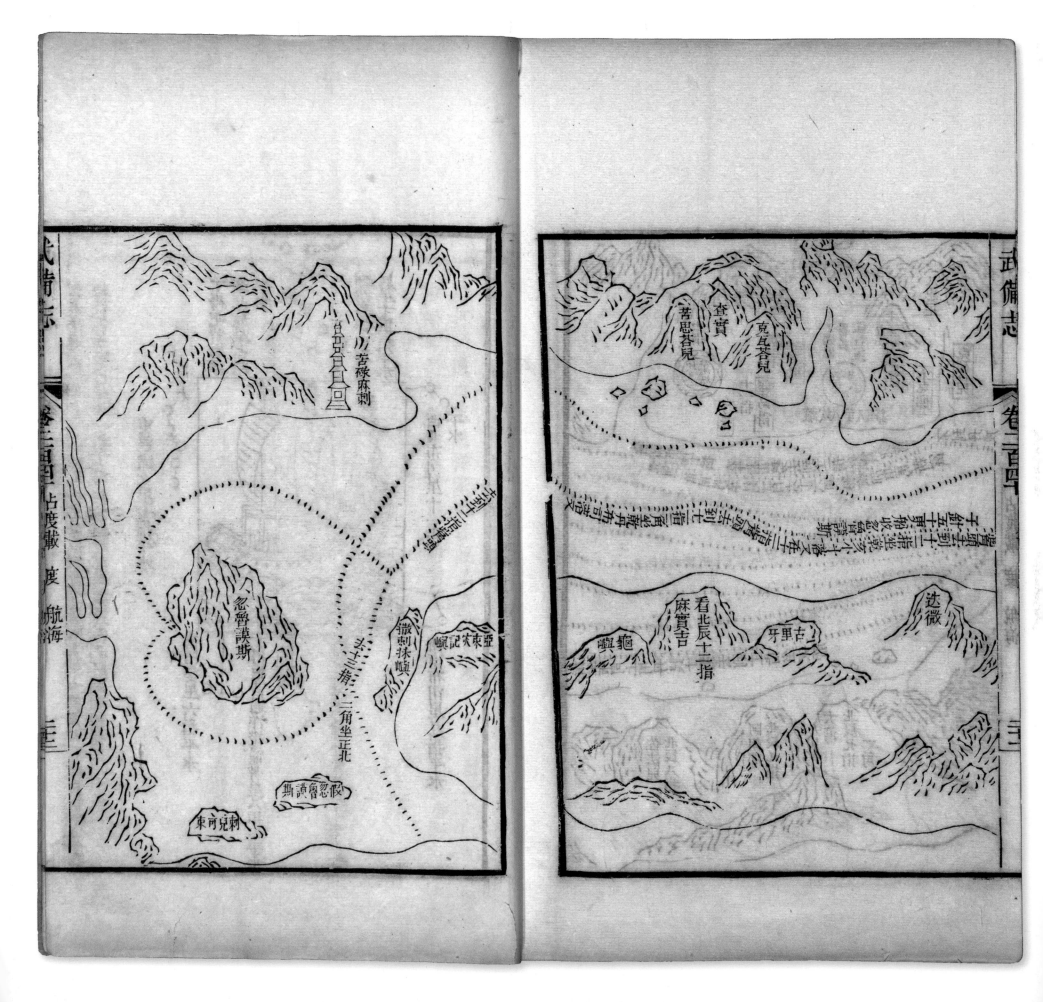

PORTOLAN CHART OF ITALY, 15TH CENTURY

Portolan charts, their name derived from the Italian for pilot books with sailing instructions, provided navigational information. This fifteenth-century chart survives in Venice which had the most impressive of the Italian maritime empires. It then included some of the Aegean islands (especially Euboea, Lemnos and Naxos), the Ionian Islands, Crete and Cyprus. However, the Ottoman advance ensured that the empire was in retreat, notably with the loss of Euboea and Lemnos. Kritovoulos, the governor of the Aegean island of Imbros, noted in 1463 that Mehmed II, the Ottoman conqueror of Constantinople in 1453, had decided to build up a fleet 'because he saw that sea power was a great thing'. This fleet swiftly became a major threat, deploying 280 galleys and other ships to capture Euboea in 1470 which undermined the Western position in the Aegean as grain from Euboea was used to supply the Knights of St John on Rhodes. Cephalonia in the Ionian Islands followed in 1479. In a war with Venice in 1499–1503, the Ottomans captured Venetian bases in Greece, notably Coron and Modon, and defeated a Venetian–French fleet at Zonchio in 1499. OPPOSITE

Vikings acted like the Polynesians who Western travellers encountered in the eighteenth century. The Vikings did so to particular effect in England, France and Ireland in the ninth and tenth centuries. There was also a major raid of Spain and Portugal in 844, although an attempt to repeat it in 859-61 was unsuccessful.

Most navigation was in familiar waters, even for the Polynesians and the Vikings, and most of this was within sight of the land. This familiarity was a matter of experience, memory and oral culture, including tales of past voyages, tales in which dangerous waters such as straits might be discussed in terms of bad spirits, as with the Strait of Messina between mainland Italy and Sicily. Acquired and accumulated knowledge was crucial, but the means to disseminate it, other than by verbal tales, were limited. That helped ensure an emphasis on older and experienced captains and sailors who had sailed particular waters before.

CHARTS AND COMPASS

In the Mediterranean, accumulated memory, passed on as verbal information, was crucial. However, there was also the development of portolan charts which supplemented sailing instructions by offering coastal outlines in order to help navigation. Their name was derived from the Italian for pilot books with sailing instructions. The charts came to be covered in rhumb lines, radiating lines showing first wind directions and later compass bearings. The charts were a guide to anchorages and sailing directions. Portolan charts became more accurate with time. New information could be incorporated into this format, while its geographical span could be expanded, as with Jacobus de Giroldis' manuscript portolan chart of 1447, which included information southwards along the Moroccan coast as far as Cape Cantin. A portolan chart of 1492

by Henricus Martellus included Bartolomeu Dias's voyage round the Cape of Good Hope.

These charts were not alone as instances of tools for navigation and thus means to record position and direction, and to assess the impact of wind and tide. The use of the compass for navigation by Westerners had begun in the twelfth century, providing a new form of information. The compass was not a one-stop change. Initially a needle floating in water, it became a pivoted indicator and, by the fifteenth century, there was compensation for the significant gap between true and magnetic north, which was an important advance. However, the capability to measure the distance travelled at sea remained an issue.

The Chinese also confronted the need to incorporate new material. The information obtained from the Chinese overseas expeditions fed into a long-established Chinese practice of assembling such material. The maritime round trip to Southeast Asia was being made frequently by the eleventh century, and nautical mapping in China originated at the latest in the thirteenth century. Zhu Siben's fourteenth-century world atlas included maps showing the Philippines, Taiwan and Indonesia. These may not seem distant, given the tendency to adopt a different scale in modern atlases when considering East Asia to that used for Europe. However, Indonesia was distant for China, while the distance from China to the Philippines compared with that across the Mediterranean. The more distant voyages of Zheng He to the Indian Ocean in the early fifteenth century, as far as the east coast of Africa, led to Ma Huan producing *Ying-yai sheng-lan* [The overall survey of the ocean's shores].

There was no resumption of such long-range activity, however, and China was central to the national world view. In 1561, a major maritime atlas

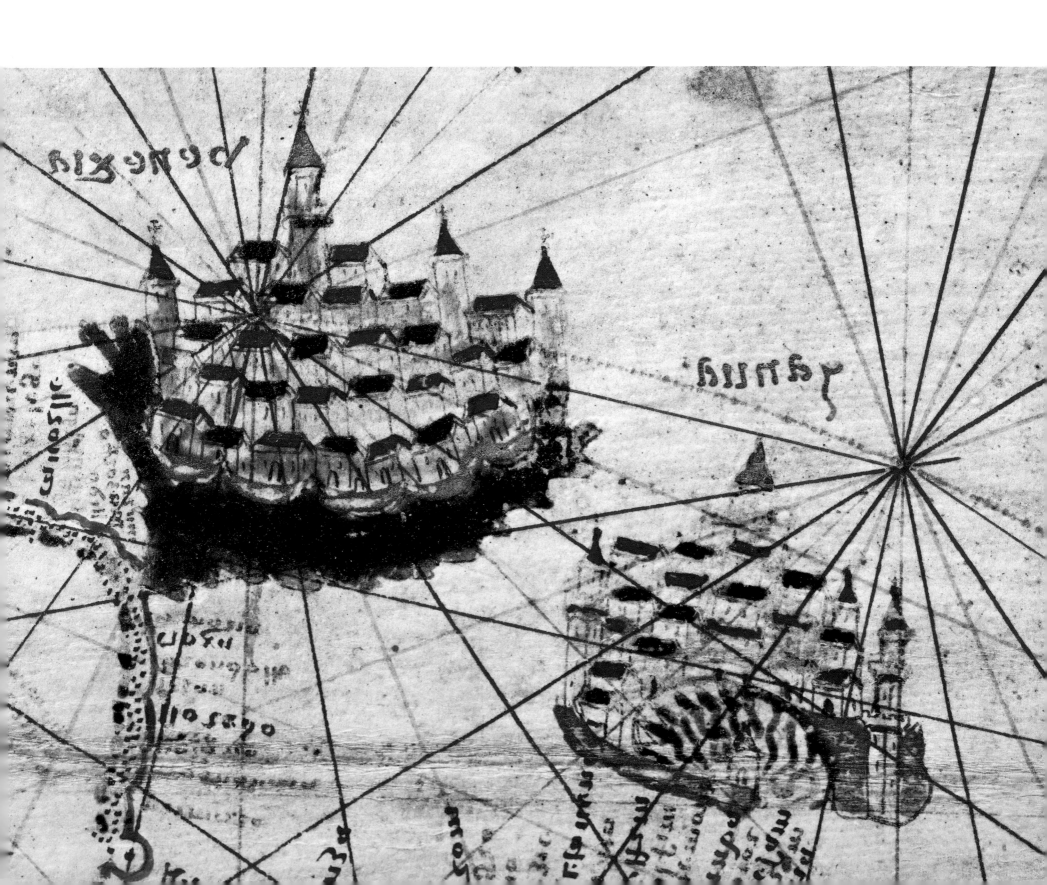

was produced in China, but it was of the country's coasts, not the world's oceans, and this at a time when the Europeans were mapping the latter. No naval maps survive for the large-scale 1590s war with Japan in Korean waters, although there were generals' maps of the coasts and of Japan that can be seen in Mao Ruizheng's *Wanli san da zheng kao* [Studies of the three great campaigns of the Wanli emperor] from 1621. There was scant equivalent elsewhere in Asia, for example in Japan. Indian navigational charts exist from the seventeenth century, but, again, they are essentially local, and certainly not global.

ISLAMIC TRADITIONS

Arab traders, benefiting from astronomical knowledge and using star compasses, had long sailed the Indian Ocean and the Mediterranean. There is some evidence that they employed charts and clearly understood the nature of the seasonally opposing monsoon winds that made navigation across the Indian Ocean predictable and, therefore, valuable for trade links. The long-standing cartographic tradition in the Islamic world was taken forward under the Ottoman (Turkish) empire, which became a major naval power in the sixteenth century, notably in the Mediterranean, but also in the Black Sea, the Red Sea, the Persian Gulf and the Indian Ocean.

The rapid Ottoman conquest of Egypt in 1517, a conquest made overland, proved particularly important in the development of naval capability and interests. Whereas the Ottomans did not have a significant naval tradition, Egypt did. Moreover, the conquest of Egypt led the Ottomans to extend their maritime interests into the Red Sea, from the base at Suez, down the coast of Arabia to Aden, which they conquered in 1538, and the coast of Africa to modern Eritrea, and from there into the Indian Ocean. Developed in the

GENOA, 1481, BY CRISTOFORO GRASSI A display of naval strength in a celebration of the recapture of Otranto from the Ottomans. Genoa was a major maritime power but, like Venice, was under pressure from the Ottomans, losing its bases of Amasra (1460) and Kaffa (1475) on the Black Sea, and Samos (1550) and Chios (1560) in the Aegean, although Corsica was retained until sold to France in 1768. Genoa focused on galley warfare and in the sixteenth century aligned with Spain, providing much of its naval power, as at the Battle of Preveza in 1538. Captured by the Ottomans in 1480, Otranto, in south-east Italy, threatened to be a base for expansion but the new sultan Bayezid II faced opposition from his brother Jem and therefore adopted a cautious international stance. Otranto was abandoned in 1481. LEFT

FRENCH RAID ON BRIGHTON, 1514

Anglo-French naval conflict in the early sixteenth century was a period of transition from medieval naval warfare, which had been dominated by coming alongside and boarding, to stand-off tactics in which warships put more of an emphasis on firepower and did not come into direct contact. In the war of 1512–14, the English and French fleets fought in the Channel in the traditional fashion, whereas off Portsmouth in 1545 they engaged in a gunnery duel. This shift had important implications for naval tactics (although truly effective ways of deploying naval firepower were not found until the next century), and it further encouraged the development of warships primarily as artillery platforms. Carvel building (the edge joining of hull planks over frames) replaced the clinker system of shipbuilding using overlapping planks, contributing to the development of stronger hulls better able to carry heavy guns. The Anglo-French War of 1512–14 saw Henry VIII support Spain, Venice and the Pope, and naval operations were a part of this wider struggle, one ultimately determined by developments in Italy. The Channel was in practice a sideshow. OPPOSITE

1510s, the Portuguese naval presence in the Red Sea was driven out. The Ottomans also extended their interests along the coast of North Africa, to Algiers in 1528 and Tunis in 1573, and thus challenged the Spanish position in the western and central Mediterranean, eventually successfully so for North Africa. A series of amphibious operations led to the consolidation of Ottoman control in the eastern Mediterranean, notably with the capture of Naxos (1566), Cyprus (1570–71) and Crete (1644–69). However, a large-scale Ottoman amphibious operation against Malta failed in 1565.

A key Ottoman figure was Piri Reis (Captain Piri), who was born Ahmed Muhiddin Piri in about 1465 in Gelibolu (Gallipoli). Having followed his uncle into the Ottoman fleet, Piri prepared both a world map of 1513 that incorporated information from Portugal and Spain and the *Kitab i-Bahriye* [Book of the Sea], a manual for sailors that provided information on routes, distances, watering locations and safe harbours around the Mediterranean. It contained more than 200 maps, including nautical charts, coastal plans and city maps, but the number varies as there were different editions, in 1521 and 1526, and numerous copies thereafter. More than thirty original manuscript copies survive.

Piri Reis was to be executed for failure against the Portuguese in the Persian Gulf in the 1550s – a reminder of the range of maritime environments in which the Europeans fought in the sixteenth century. The Portuguese, in particular, encountered a wide range of regional naval powers: notably China in the 1520s, Calicut and Gujarat in India in the 1490s to 1500s and 1530s respectively, Mameluke Egypt in the 1510s, and, later, the Ottoman Empire, as well as Atceh and other regional powers in the East Indies. In contrast, Spain's opponents in the New World, the Aztecs, Incas and Mayas, did not have navies,

although in the Philippines Spain did encounter naval opposition.

GLOBAL RANGE AND THE NEED FOR INFORMATION

European navigators had an unprecedented maritime range from the late fifteenth century, reaching the Indian Ocean round Africa and crossing the Atlantic. In 1520–21, Ferdinand Magellan was the first known individual to cross the Pacific (and probably the first to do so), and map-makers stored, incorporated and reproduced the resulting information. Although Magellan was killed on Cebu in the Philippines before it could be completed, the first circumnavigation of the world, completed by Juan Sebastian del Cano in 1522, arose from this expedition, and this circumnavigation made the globe an obvious tool for understanding the world. Furthermore, new information clarified the amount of yet more information that had to be acquired. The globe, and the graticule that covered it, had to be filled. What filled it in maps was often drawings. These classically displayed large fish, but the marine creatures frequently shown included imaginary ones, and they were sometimes in conflict with ships, attempting to drag them to destruction.

The Mercator projection, published in 1569, offered a response to the problems of the depiction in two dimensions of the earth, a sphere, by providing negligible distortion on large-scale detailed maps of small areas. Gerardus Mercator (1512–94), who never went to sea, looked back to Ptolemy in employing coordinate geometry as a guarantee and means of a mathematically consistent plan and logically uniform set of rules. The combination of the grid of latitude and longitude with perspective geometry proved a more effective way than portolan charts to locate

places, and thus to adapt to the range of new information. Portolan charts had similar or identical chart symbols but lacked common scales and units of measure, and were essentially directional guides based on analogue, rather than digital, methods.

This was the intellectual background to the production of maps for a European public. There are some good examples of earlier maps produced by woodblock printing, and improvements in printing technology ensured that the public could be more readily served. The key early centres of European map printing were all foci of maritime trade, notably Venice, Antwerp and Amsterdam. As a result, they acted as clearing houses for new as well as established information. Printed portolan charts came to replace hand-drawn ones. Amsterdam, in particular, was a commercial centre rather than a typical military one, and the motivation for maps there was commercial rather than military. The commercial map-maker/producer made his living by creating the best possible maps. Without a monopoly, this led to an improvement in the maps available.

Vasco da Gama arrived in Indian waters in 1498 carrying cannon that less heavily gunned Asian warships could not resist successfully in battle. This technological gap helped give the Portuguese victory over the Calicut fleet in 1503, with Portuguese gunfire seeing off boarding attempts. The Portuguese were also successful over other Indian fleets, those of Japara and Gujarat, in 1513 and 1528 respectively. An Egyptian fleet sent from Suez in 1507, and supported by Gujarati vessels, was initially successful against the Portuguese at Chaul in 1508 but largely destroyed off Dill in 1509. The standard armament of a Portuguese galleon in 1518 was thirty five guns. The Portuguese also seized positions, notably Goa in 1510 and Malacca in 1511, although deep-draught Portuguese warships frequently had only limited value in shallower waters inshore, including those in estuaries and up rivers. In 1512, a base was established at Bantam in the Sunda Strait between Java and Sumatra and in 1518 bases were founded at Colombo and Galle in Sri Lanka.

OPPOSITE

REASONS FOR NAVAL MAPS

In the centuries prior to 1700, naval battles were less frequent than those on land and were commonly celebrated pictorially on canvas or fresco, and generally with a highly dramatic representation focusing on close-quarter fighting in the shape of boarding. Nevertheless, there was interest in maritime maps as an aspect of this illustrative record. In 1590, popular topicality was displayed when the publication of an English translation of an account of the defeat of the Spanish Armada two years earlier included a set of eleven maps by Robert Adams showing the successive stages of the campaign on a background of large-scale maps of the English Channel. This proved a clear guide to the setting of the engagements.

The presentation of the Armada (as of the major defeat of the Ottoman fleet by a Spanish-Venetian-Papal fleet at Lepanto in 1571, the largest ever and most dramatic battle in the Mediterranean) was one of the key aspects of mapping naval conflict, that of displaying what had happened for reasons of glory and/or profit. In Western Europe, the depiction of naval conflict became more frequent in the seventeenth century, most commonly using illustrations but also increasingly maps. Etchings, engravings or a mix were often published as a stand-alone broadsheet, sometimes including a block of text. In the United Provinces (Dutch Republic), these broadsheets became more common in the 1630s and 1640s. The Dutch defeat of the Spanish fleet at the Battle of the Downs in 1639, Marten van Tromp beating a larger fleet under Antonio de Oquendo, was a major blow to Spanish power in the Low Countries which. This battle, which forced Spain to rely on land routes to its possession of the Spanish Netherlands (modern Belgium), was marked in many Dutch maps and illustrations. Such maps were often engraved and published by specialists in the news world, rather than by cartographers. This was presumably because the sale of these broadsheets required networks in the current affairs trade. A map or illustration were fitting memorabilia after a decisive victory.

Another reason for maps was that of assisting military planning, although the process of doing so is shrouded in obscurity. In the months prior to the Spanish Armada, Anthony Ashley, secretary to the English Privy Council, sponsored the translation into English of Dutch charts by Lucas Janszoon Waghenaer covering the seas from Cadiz to the Netherlands which had appeared as the *Spieghel der Zeevaerdt* in 1584–85. The English edition appeared as *The Mariner's Mirrour*. French and German editions also appeared. Subsequently, established chart publishers were called on in the seventeenth and eighteenth centuries to meet official needs for charts, for example, the Blaeu dynasty in connection with the Dutch East and West India Companies in the seventeenth century and the publication of the *Neptune François* and the Dutch pirated version in the early 1690s. The Dutch played an important role not only in gathering new information but also in circulating material previously available only in manuscript or in printed versions that had not circulated so widely, such as the Italian portolan charts used by William Barentsz.

In a direct and ultimately successful challenge to Dutch hegemony over the sea chart market, a market crucial for mariners, the first part of *The English pilot* by John Seller was published in London in 1671, with the last part, part five, appearing in 1701. Encompassing the known globe, numerous editions were published throughout the eighteenth century. Although many of the maps were derivatives of earlier Dutch charts, the work proved enormously popular, not least due to the charts' accompaniment with

CONSTANTINOPLE AND SCUTARI, 1521, FROM
KITAB-I BAHRIYYE BY PIRI REIS This is a map
of Constantinople (right) and Scutari
(left), with north shown at the bottom
of the map and the Golden Horn on
the right hand side. Naval power was
important to the expansion of Ottoman
power, including in the capture of
Constantinople in 1453. The Ottomans
moved ships overland to the Golden
Horn from which waterway they could
put further pressure on the defenders.
Thereafter, the Ottomans developed a
major fleet based at the naval arsenal at
Galata, on the other side of the Golden
Horn to Constantinople. Expanded under
Selim I (r. 1512–20), whose conquest of
Egypt in 1517 was helped by his fleet,
Galata became a major centre of
shipbuilding, notably in the rapid
construction of a new fleet, including
150 galleys, after defeat at Lepanto
in 1571. OPPOSITE

sailing directions written in English. English knowledge of foreign languages was limited. The relationships between communication links, linguistic knowledge, literacy, literature and publication greatly helped the Dutch map-makers of the time.

Dutch maps were the first port of call for the Baltic Sea, the key area for trade in northern Europe and one that was especially significant as it was from there that Europe was supplied with naval stores (timber, hemp, flax), grain and iron. The Swedish navy used Dutch maps of the Baltic Sea until 1644 when the first Swedish map was published by the navy's senior master pilot, Johan Månsson. This map was heavily dependent on the Dutch maps. A much better map was published by Petter Gedda in 1694 after extensive surveying operations. In 1687, Gedda had been appointed the first director of the Navigation Bureau established under the Admiralty to produce charts. From then on, the Swedish navy continually surveyed and updated maps of the Baltic Sea and the Swedish coast. It needed to do so since naval links were crucial to Sweden's ability to maintain its territorial position on the eastern and southern shores of the Baltic Sea and to move troops in response to threats, as Charles XII did in 1700 when defeating the Russian army under Peter the Great in the Battle of Narva. Information was a capability enhancer.

The sense of mapping as a vital aid to national defence was seen in England in 1681 when the government appointed a naval officer, Captain Greenville Collins (*c.* 1634–94), commander of the eight-gun yacht HMS *Merlin*, 'to make a survey of the sea coasts of the kingdom by measuring all the sea coasts with a chain and taking all the bearings of all the headlands with their exact latitudes'. Collins had extensive experience of navigation, including a failed attempt to reach Japan by a north-east passage north of

Asia, which ended with the vessel wrecked off Novaya Zemlya. He had also served in the Mediterranean against Algerine privateers (pirates to the English), who were a major threat to trade, and drew maps on his mission. Collins lobbied for an improved survey of Britain's coast, both to rectify mistakes and to provide a centralised system for collecting and disseminating improved maps. The survey, which lasted seven years, had many problems due to the speed with which it was accomplished, the limited manpower available and the lack of an available comprehensive land survey of the coastline as the basis for a marine survey. More generally, maritime mapping is in part dependent on its land counterpart, both practically in fixing the coastline and what is on it and functionally in terms of linking to amphibious operations.

Collins published his results in *Great Britain's Coasting Pilot* which contained sailing directions, tide tables, coastal views and charts. The complete work was first published in 1693. Collins was allowed to style himself Hydrographer in Ordinary to the King from 1683. The survey was reprinted frequently in the eighteenth century. The interest in such mapping was utilitarian. In 1670, Charles II ordered the Council of Plantations 'to procure maps or charts of all … our plantations abroad, together with the maps … of their respective ports, forts, bays and rivers'.

SECRECY

One testimony to the value of maps was provided by the many attempts to restrict their availability. These attempts underlay the extent to which non-Western governments did not publish much material. This tendency was not restricted to non-Western powers. In particular, the Spanish government, which had sought to keep secret information about its discoveries from the outset, came to focus on trying to keep other

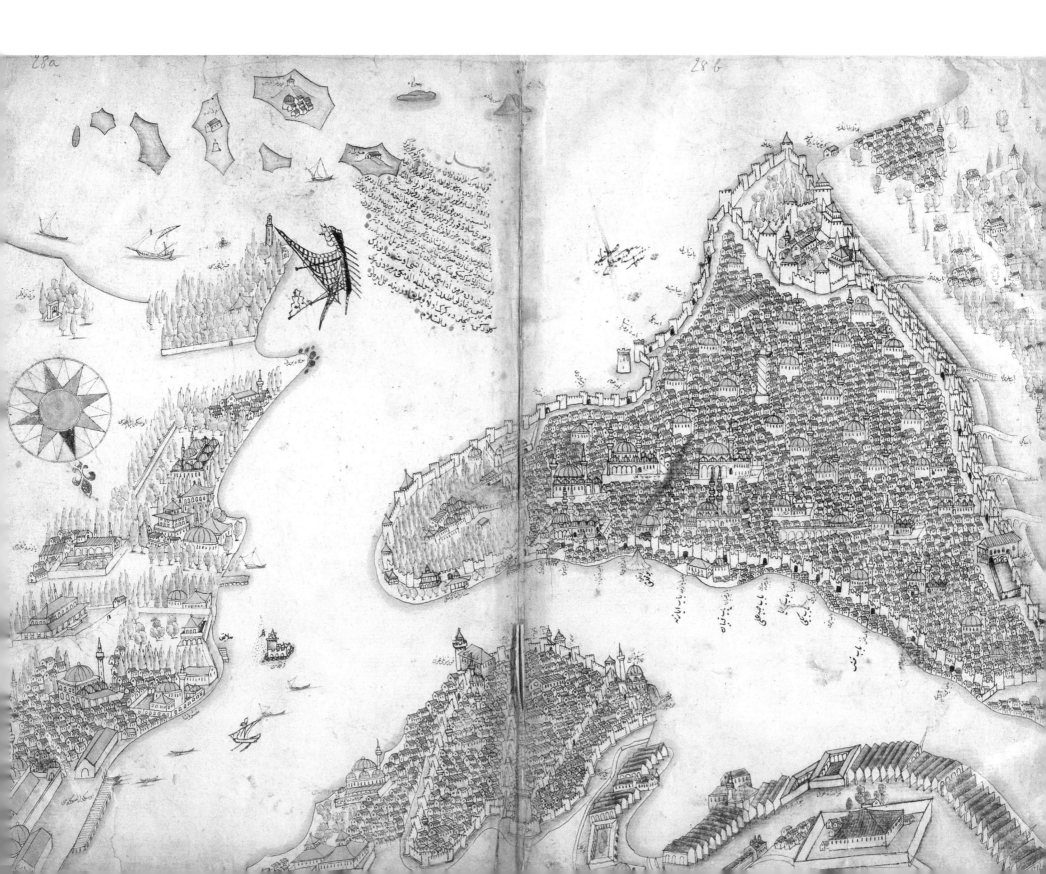

CHINESE DEFENSIVE POSITIONS, 1524–26 *Wan li hai fang tu shuo* (Illustrated map of Qing Empire coastal fortifications), originally drawn during Ming Jiaqing period between 1524 and 1526. This map is a copy drawn in 1725 and shows the coastline including defensive positions. Chinese concern increased greatly due to the arrival of the Portuguese. In 1521, the Chinese clashed with Portuguese ships off T'un-men, near Macao, when they tried to force an expulsion of all foreigners, after the death of the empreror, Cheng Te. The outnumbered Portuguese were put under heavy pressure with the Chinese attacks beaten off, but with difficulty and with one ship sunk. In 1522, the Portuguese lost two more ships to a Chinese squadron employing cannon. Thereafter, until 1528, Chinese fleets were deployed each year to prevent any Portuguese return. **BELOW**

Europeans out of the Pacific, which it regarded as a monopoly, and on restricting information about the ocean.

However, in 1680 a band of English buccaneers under Bartholomew Sharp crossed the Isthmus of Darien from the Atlantic to the Pacific not far from the present Panama Canal. Using a Spanish ship they had seized off Panama, they attacked Spanish shipping, before returning to England in 1682. The band included Basil Ringrose, who both wrote a journal of the expedition that was published in 1685 and compiled a substantial 'waggoner' – the term then used for a description in the form of sailing directions – to much of the coast he sailed along, as well as to some parts he never visited. This description stemmed from the *derrotero* (set of official manuscript sailing directions), illustrated by a large number of coastal charts, that Sharp seized from a captured Spanish ship in 1681 and presented to Charles II in order to win royal favour. Such atlases had been regarded by the Spaniards as too confidential to go into print, which therefore limited their circulation and meant that the standardisation, for good or ill, offered by print did not occur.

Earlier, in the 1590s, Jan Huygen van Linschoten, who had worked as a clerk and scribe for the

Portuguese Archbishop of Goa in India and been involved in map-making, returned to the Low Countries and passed on the cartographic information that the Portuguese had striven to keep a secret. This helped the Dutch East India Company to deploy its warships in the Indian Ocean against the Portuguese. In a series of amphibious operations, the Dutch drove the Portuguese from Malacca, coastal Sri Lanka and some positions in India. The Dutch positions in regional trade rested on naval strength.

SIMILARITIES TO MODERN NAVAL WARFARE

Naval warfare was scarcely a matter of the crump of battleship guns or the whoosh of jets taking off from carriers. Nevertheless, in an age of sail or oared power, there were many characteristics similar to those of modern naval operations. First, despite the examples frequently noted, battle was only occasional and most naval operations did not entail the search for battle. Weaker forces sought to avoid battle. Secondly, many operations were in concert with troops. Amphibious power was a key element. This helped ensure that the most significant maritime environments were coastal waters. They were the prime areas of trade and fishing, since much of the population lived in coastal regions.

Maritime force at its most significant tended to involve attempts to overcome foes by mounting invasions, as in the cases of the unsuccessful Persian attacks on Greece in 490 and 480 BCE (the latter of which led to a major naval defeat at Salamis), the Norman invasion of England in 1066 CE and the unsuccessful Mongol attacks on Japan in 1274 and 1281. Operations across nearby waters were the key element as it was necessary to supply and support large forces and to protect them from the risk of storms as well as from becoming becalmed. An invasion at a great distance was not easy. Far from this characteristic being a feature of the pre-steamship age, it could also be seen with the Allied invasion of German-held Normandy in 1944, a crucial event in that war. However, two years earlier, Operation Torch, the American-led invasion of north-west Africa, provided an excellent instance of a long-distance invasion as much of the invasion fleet had travelled across the Atlantic.

Effectiveness was in large part a matter of the factors, both military and, crucially, political, that led to success for the troops once landed. This was notably the case with William III of Orange's successful invasion of England in 1688. Its success initially rested on the failure of the English fleet to intercept the Dutch, a case of a decisive non-battle in which luck played a major role. Finally, success rested on political willingness in England to accept the new order. James II deployed a larger army to block William's advance on London, but this force rapidly dissolved due to his collapse of will, a collapse that owed much to key betrayals.

At the same time, there were the questions to do with naval power, as applied to the likely role of the ship in question. Questions of sea-worthiness, capacity, draught, speed, armament and crew all played a role. The balance varied greatly depending on the waters and tasks in question. Cannon became increasingly important in naval armament in the fifteenth century and, even more, in the sixteenth century, but not all warships had or relied on stand-off firepower or on such firepower alone. This reliance was particularly so for the hull-mounted cannon of deep-draught warships.

This variety underlines the need not to assume a single definition of naval power, nor to assume that there was only one direction of development, that towards the modern day. If deep-draught warships were indeed dominant in the nineteenth century,

ALGIERS EXPEDITION, 1541, FROM *CIVITATES ORBIS TERRARUM* BY GEORG BRAUN AND FRANS HOGENBERG Spain seized the Peñón d'Argel position dominating Algiers in 1510, but in 1529 it was recaptured by Hayreddin 'Barbarossa', a notable corsair (privateer) whose submission to the Ottoman sultan, Süleyman I, was rewarded in 1533 when he was appointed *Kapudan Pasha*, the admiral to the Mediterranean fleet. Süleyman became a major challenge to Habsburg power, which led the Emperor Charles V to launch a major expedition against Algiers in 1541. This was a large-scale amphibious expedition involving sixty-five galleys, 450 support vessels and 34,000 troops. However, while landing the troops, the fleet was badly damaged by an autumnal storm and, with about 150 ships lost, the troops were soon re-embarked. There was no repetition of Charles' success against Tunis in 1535. In the 1550s, the Ottomans sent large fleets into the western Mediterranean, but the Ottoman naval presence there was largely a matter of Algiers-based squadrons. A major privateering base, Algiers was frequently attacked, including unsuccessfully by Spain in 1775 and 1784, finally falling to the French in 1830. OPPOSITE

that was not the case in the sixteenth when galleys were particularly important. This was especially so in the Mediterranean but was also in northern European waters, for example in the Gulf of Finland and off western Scotland, as well as in Southeast Asia where the navies of Atceh and Sulawesi depended heavily on galleys. Galley warfare focused on closing and boarding which, more generally, was an important alternative, or sequel, to stand-off firepower. Galleys carried cannon, but much of their space was devoted to rowers and their oars, which restricted the possible use of cannon. Deep-draught warships faced serious problems in inshore waters. This, for example, was an issue for the Portuguese in West Africa. Adapting to such waters, the Portuguese used galleys in some areas, notably the Bay of Bengal.

Whereas most Eurasian empires that had a coastline focused on landed élites, yields from land and land warfare, a profitable co-operation of naval power and trade became the characteristic of European maritime states. This was the case with the most successful: Venice in the sixteenth century, the Dutch in the seventeenth and Britain in the eighteenth. Other countries failed to match this synergy and, in particular, lacked a focus on long-range naval activity. As a result, naval warfare largely became the history of the European maritime states.

The major effort that China, Japan and Korea put into naval conflict in the 1590s, with China and Korea opposing a large-scale Japanese invasion of Korea, eventually successfully so, was not sustained. Japan ceased to mount invasions, while Chinese naval power was devoted to local tasks, notably opposition to piracy, a role that eventually culminated in an invasion of Formosa (Taiwan) in 1683. Earlier, a pirate leader had invaded Formosa and captured the Dutch base. The Manchu or Qing dynasty that conquered China

THE PACIFIC WITH MAGELLAN'S SHIP FIRING A CANNON, FROM *THEATRUM ORBIS TERRARUM* **BY ABRAHAM ORTELIUS, 1570** Although Ferdinand Magellan was killed at Cebu in the Philippines in 1521, his expedition in the service of Spain in 1519–22 was the first to circumnavigate the world, a dramatic achievement that established only a flimsy presence but one that was more potent because no other Pacific polity deployed significant naval power able to span the ocean. In particular, China and Japan had not developed their potential. In the early 1560s, Spain established a colony in the Philippines that traded with Mexico and, from there, Spanish bases were founded in the western Pacific. However, the Spaniards encountered serious resistance from the areas in the Philippines where Islam had made an impact: the southern islands of Mindinao and Sulu. China and Japan made limited attempts to drive Spain from the western Pacific, but with scant success. In 1574, an attack on Manila by Lin Feng, a Chinese pirate, was driven off. OPPOSITE

in the 1640s and ruled it until 1912 focused on land warfare. When the Qianlong Emperor attacked Burma in the 1760s, the invasion was mounted overland and not supported by naval action. Similarly, the attack on Vietnam in 1788 was overland.

NAVAL CONFLICT AT ITS PEAK

The major naval battles of the sixteenth century, notably Lepanto and the Armada, acquired totemic significance and were extensively memorialised. However, in practice, naval conflict in European waters reached a high point in the second half of the seventeenth century, without there being any equivalent elsewhere in the world. There had been notable naval battles in the first half such as the Dutch defeat of the Spaniards in the Battle of the Downs in 1639. The war between France and Spain from 1635 to 1659 also had a naval component, mostly in the form of amphibious actions in the western Mediterranean.

Such battles became more common from 1652 to 1692 as England, the Dutch and France vied for naval superiority and built warships and developed supporting infrastructures accordingly. Specialised bases were developed, including Brest for France, Portsmouth for England, Karlskrona for Sweden and Copenhagen for Denmark. These bases housed docks and major storage yards of timber, sail and cannon. The repair facilities had to be good as ships were made of organic materials that rotted.

Naval superiority was designed to serve strategic goals, permitting or preventing invasion, for example, of England and Sicily. Moreover, naval strength was regarded as crucial for trade protection and as important for prestige.

There was a comparable rivalry in the Baltic Sea involving Denmark and Sweden to those involving the Western European powers. The Baltic rivalry attracted less attention, in part because it did not extend as widely as those involving the Western European powers, but Baltic conflict contributed to the sense of naval warfare as normal and to be prepared for as a matter of routine. There was a particular requirement for accurate mapping due to the number of offshore islands and rocks.

The frequency and nature of battles encouraged illustration. The exchange of fire between warships and the capture or destruction of particular vessels were of the greatest prominence. At the same time, there was a stronger interest in maps in order to place the action. The three Anglo-Dutch wars between 1652 and 1674 encouraged this process as the two powers were the major centres of map-making. With larger ships, the English fleet proved stronger than the Dutch in the first war, that of 1652–54. In contrast, in the second, that of 1665–67, the English were handicapped by French support for the Dutch, while the Dutch were strengthened by a major shipbuilding programme in which they built warships as large as those of the English. The intensity of battle was high, with frequent clashes, including the successful Dutch raid on the English warships laid up in the River Medway due to financial problems. In the third war, that of 1672–74, the Dutch confronted both England and France and fought a successful defensive war, not least because they were able to deny their opponents the possibility of invasion by sea. These conflicts saw the development of line-ahead tactics for warships as a means both to maximise firepower and to provide cohesion. The distinction between armed merchantmen and warships became clear, total firepower increased and navies came to take the form seen throughout the remainder of the age of sail.

MARIS PACIFICI,
(quod vulgò Mar del Zur)
cum regionibus circumiacentibus, insulisque in eodem
passim sparsis, novissima descriptio.

AMERICAE SEPTEM=
TRIONALIOR PARS.

MARIS ATLANTICI

SIVE MAR DEL NORT

Noua Hispania.

PARS.

Spagnola

MARE PACIFI=

CVM, QVOD VVLGO

Caribana.

NON=VNT

AMERICÆ
MERIDIO=
LIOR PARS

Peru

Charcas

Nova Guinea, quibusdam
Terra de Piccinacoli.

Insulæ Sa=
lomonis.

Circulus Capricorni.

Chili.

MAR

DEL

SPE ET
METV.

Prima ego velivolis ambivi cursibus Orbem,
Magellane novo te duce ducta freto.
Ambivi, meritoq vocor VICTORIA: sunt mi
Vela, alæ; precium, gloria; pugna, mare.

ZVR

GENIO ET INGENIO NO-
BILI DN. NICOLAO ROCCOXIO,
PATRICIO ANTVERPIENSI,
EIVSDEMQVE VRBIS SENATORI,
Abrahamus Ortelius Regiæ Mtis geographus
lub. merito dedicabat.

15 89

TERRA AVSTRALIS,

SIVE MAGELLANICA, NON=

DVM DETECTA.

Cum privilegiis Imp. & Reg. Maiestatum,
nec non Cancellariæ Brabantiæ, ad decennium.

Patago

Archipe=
lagus in=
sularum.

Fretum Magella=
nicum

Tierra del Fuego

LEPANTO, 1571, BY HORATIO DE MARII TIGRINO (RIGHT) AND ROME, 1572 BY IGNAZIO DANTI (1536-1586) In response to the Ottoman invasion of Cyprus in 1570, a Holy League of Spain, Venice and the Papacy deployed a fleet under Philip II's illegitimate half-brother, Don John of Austria. It found the Ottoman fleet, under Müezzinzade Ali Pasha, at Lepanto off the west coast of Greece. Don John had 236 ships and 1,815 cannon, the disease-ridden Ottomans about 230 and 750. More than 100,000 men took part in the battle of 7 October 1571. The willingness of both sides to engage in an open sea battle was crucial, as the Ottomans could have pulled back under the guns of the fortress at Lepanto and forced the Christian forces into a risky amphibious assault. Good morale and determined leadership characterised both sides. Don John relied on battering his way to victory, although he also benefited from having a reserve squadron, which permitted a response to the success of the Ottoman offshore squadron. Superior Christian gunnery, the fighting qualities and firepower of the Spanish infantry, who served on both the Spanish and the Venetian ships, and the exhaustion of Ottoman gunpowder all helped to bring a crushing victory in four hours' fighting. The cannon of six Venetian galleases played an especially important role in disrupting the Ottoman fleet. **RIGHT AND FAR RIGHT**

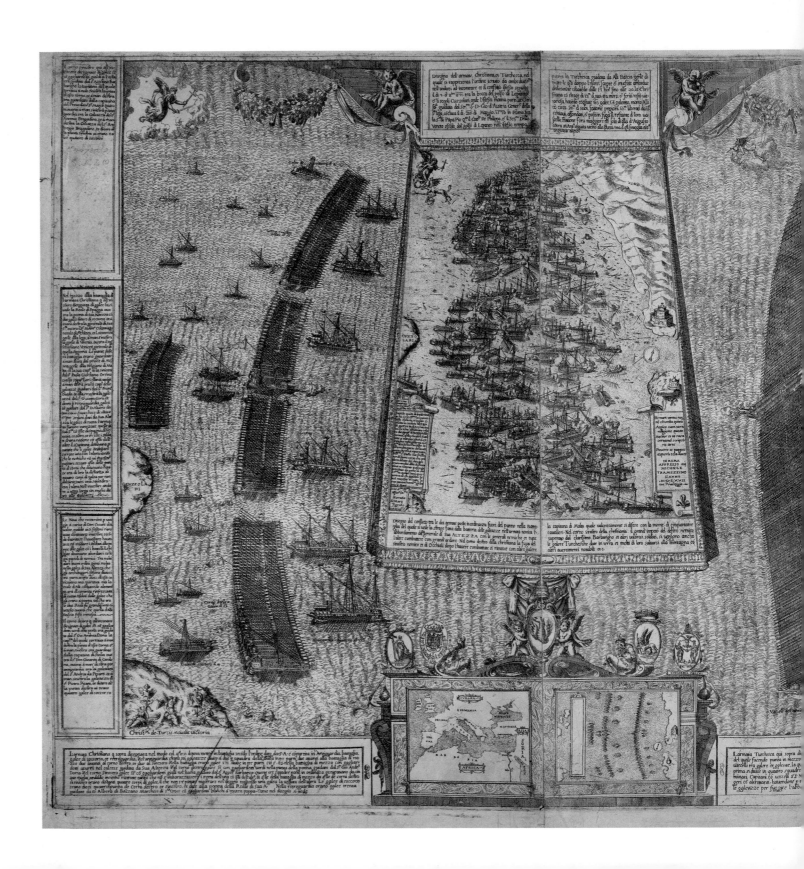

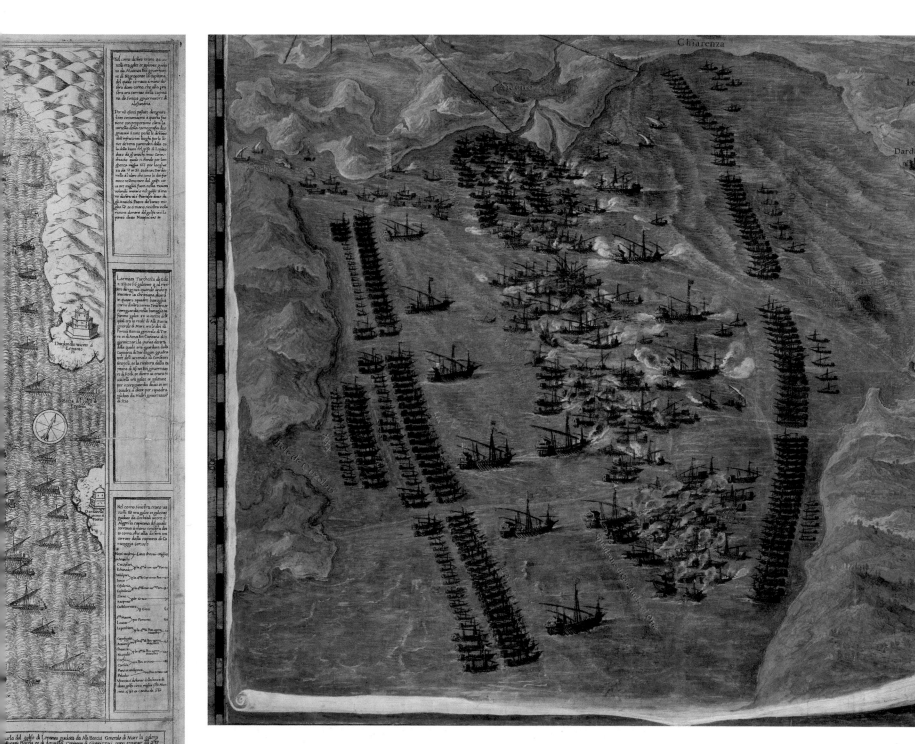

**DRAKE'S ATTACK ON CADIZ, 1587, BY WILLIAM
BOROUGH** Elizabeth I of England and Philip
II of Spain formally went to war in 1585,
ensuring that Spain had an enemy that
could only be decisively defeated at sea.
Spain could deploy significant naval
strength before, as in the conquest of
Portugal in 1580 and then in 1582–83 off
the Azores, in which an opposing French
fleet was defeated at Ponta Delgada
in 1582. The Spanish fleets used a
combination of galleons and galleys, but
the planned invasion of England was of a
totally new order of magnitude for Spain
and for Atlantic expeditions, and its
scale and ambition helped mark a major
extension in naval operations. It was
postponed because of the English
spoiling attack under Sir Francis Drake on
the key Spanish naval base at Cadiz. The
lack of reconnaissance capabilities made
surprise attacks possible. Borough, the
more cautious Vice-Admiral, complained
that Drake had conducted his command
in an autocratic fashion, which led to
Drake placing him under arrest. **RIGHT**

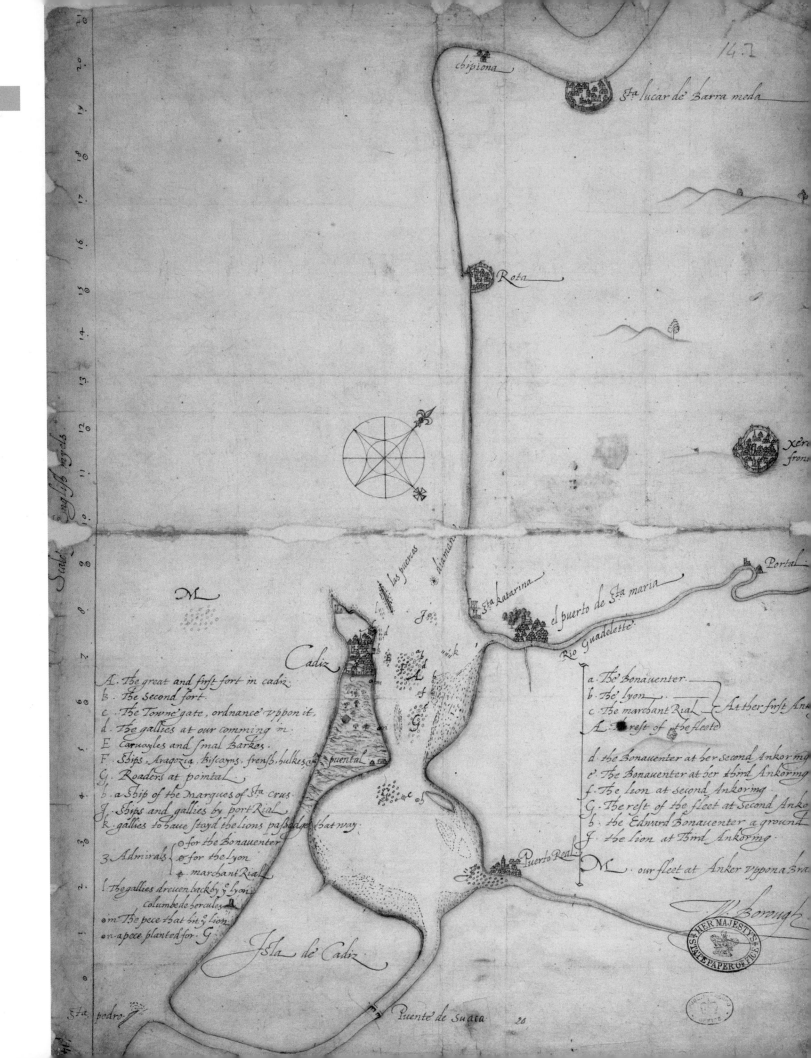

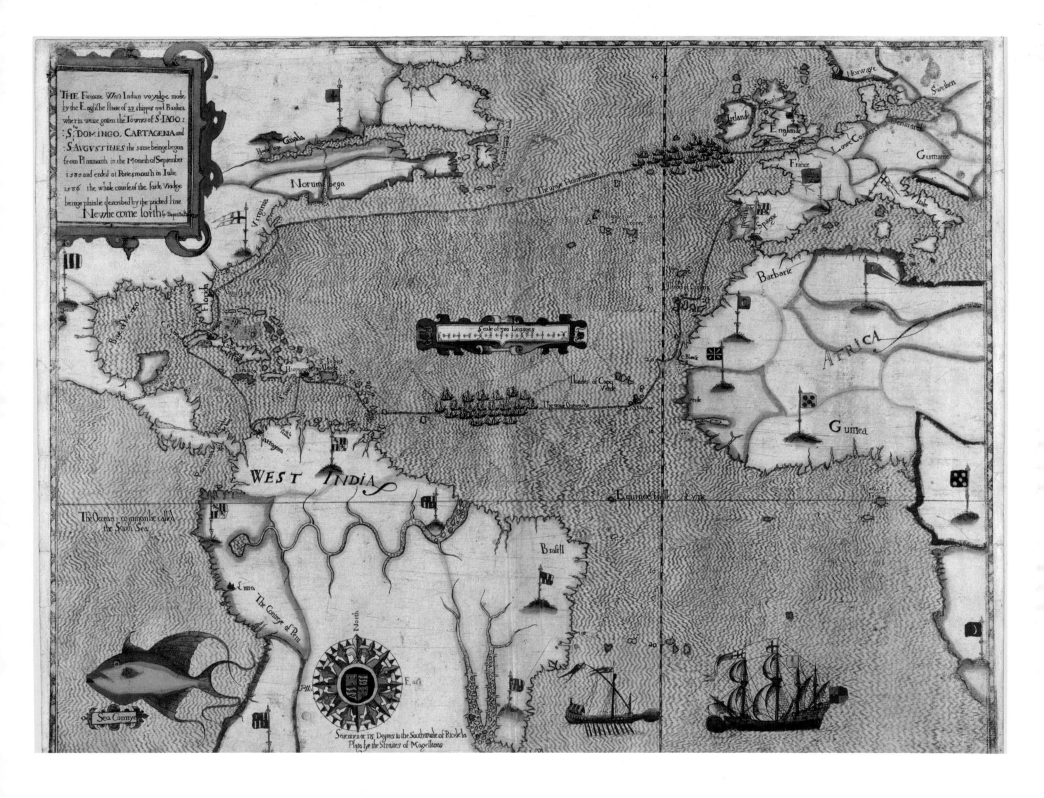

MAPS OF DRAKE'S VOYAGE IN 1585–86, FROM A
SUMMARIE AND TRUE DISCOURSE OF SIR FRANCIS
DRAKE'S WEST INDIAN VOYAGE, BY BIGGES AND
CROFTES, 1589 Commanded by Sir Francis
Drake, the fleet sailed from Plymouth on
14 September 1585. It raided the Spanish
port of Vigo en route to the Caribbean
island of Dominica, then captured San
Domingo in Hispaniola on 1 January 1586
and went on to Cartagena which was
occupied and ransomed. The English fleet
lost many men to sickness and travelled
home via Florida, where Spanish positions
were raided, and then to North Carolina,
returning home with the colonists and
reaching Portsmouth on 28 July. **ABOVE**

SPANISH ARMADA, 1588, BY ROBERT ADAMS,

PUBLISHED IN 1590 The Spanish invasion plan required the fleet sailing from Spain through the English Channel to provide cover for an invasion by the large Spanish army in the Low Countries. The assumption that such coordination could be obtained was highly optimistic, although the Spanish army had access to ports from which an invasion could be mounted. The English harassed the Spanish fleet in the Channel, but were unable with their long-range gunnery to inflict any serious damage on the highly disciplined sailing formation that the Armada adopted. With the advantage of superior sailing qualities and compact four-wheeled gun-carriages, which allowed a high rate of fire, the English fleet suffered even slighter damage, although it was threatened by a shortage of ammunition. Many of the Spanish guns were on cumbersome carriages designed for use on land. This map shows two stages of the combat and captures the discipline of the Spanish formation.

ABOVE

THE ATTEMPTED SPANISH INVASION OF ENGLAND, 1588, BY ROBERT ADAMS, PUBLISHED IN 1590

When the Spanish fleet anchored off Calais, it found that the army had been able to assemble the transport vessels necessary to embark for England but could not come out until after the English and Dutch blockading squadrons had been defeated. The Spanish fleet, however, was badly disrupted by an English night attack using fireships. Coming to close range to engage in a mêlée using its cannon and muskets, the English fleet then inflicted considerable damage in a running battle off Gravelines, driving a number of the Spanish ships onto the sandbanks. The brunt of the battle was borne by the Portuguese galleons in Spanish service as they were experienced in stand-off gunnery. A strong south-westerly wind drove the Armada into the North Sea, whence it returned to Spain via the hazardous north-about route around the British Isles suffering heavy storm damage. LEFT

SIEGE OF ENNISKILLEN CASTLE, 1594 Rowed boats continued to be important in conflict in the Western Isles of Scotland and northern Ireland in the sixteenth century, with the long-established Viking tradition of longship building continuing in the Isles. Galleys were also used by the English, as here at Enniskillen. Conflict in Ireland in the 1590s demonstrated the strength of the Irish combination of traditional irregular tactical methods with modern firearms, but also the advantages the English had when up against defensive positions. RIGHT

The greate bote wᵗʰ 67 men for breache ar going

Eneſkillin Caſtell.

The ſcaling.

which is paſes att ſ ſcott the paſe. this Caſtell
Taken the 1/ of ſſebruare 1593: by
Cayttn John dowdall then governor
Made and dnn by John Thomas
Solder.

Captaine Bingham's Cattle.

Pawkon

Robonet

Muſketeres

Captaine George Bingham

Captaine Binghams Campe.

A doble ditche deepe water.

Breach

The bote ankered to breach.

Magwiers hote.

Muſketers

ſowe men

A quadrant to ſecon yf neede

Captaine Dowdull governor

Three falcons

Muſketeres

The Paſſage to bellyke.

A dyche cutt.

An Iſlande.

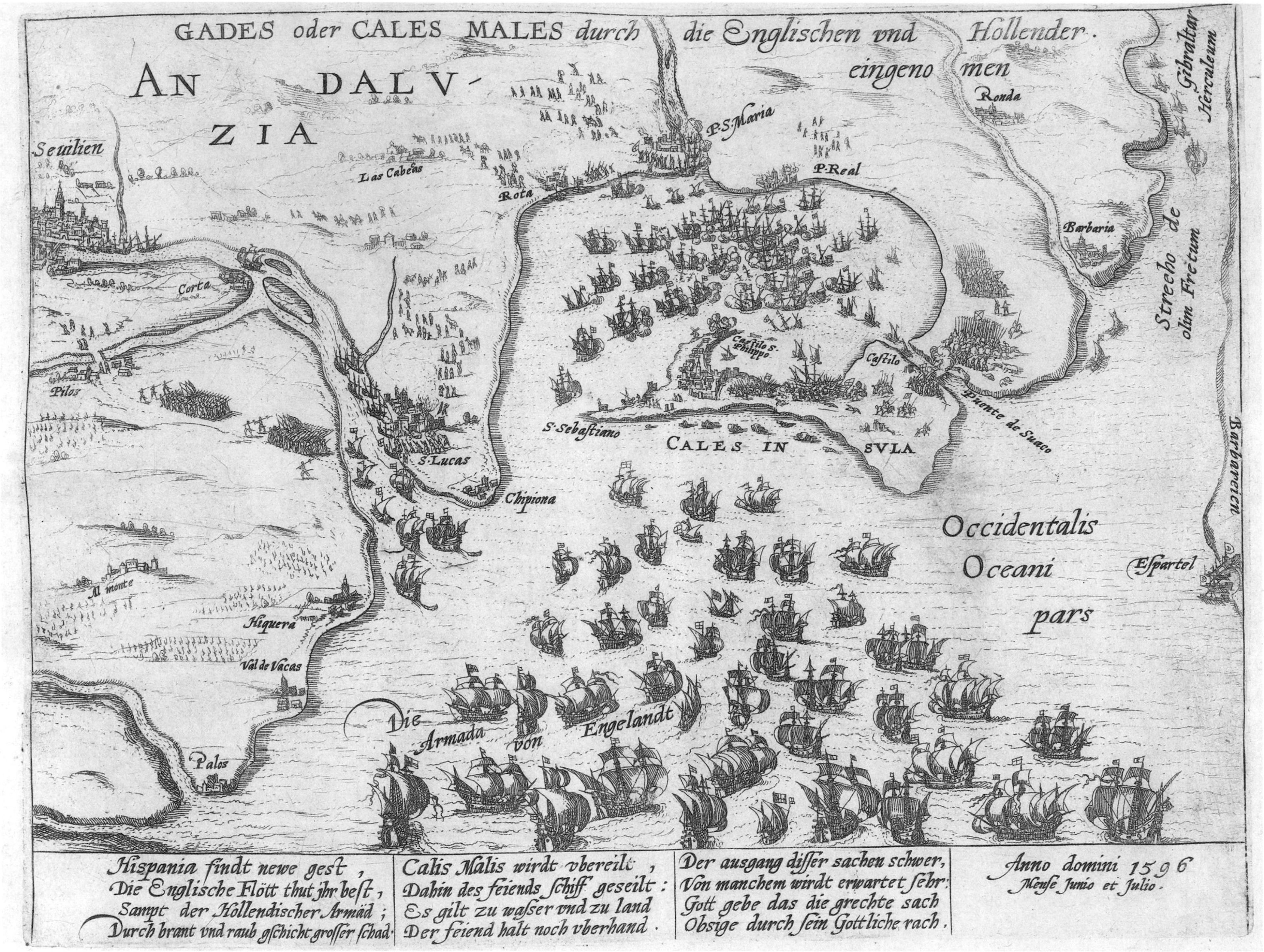
GADES oder CALES MALES durch die Englischen vnd Hollender. eingenommen
AN DALV ZIA
Seuilien
Las Cabeſas
Rota
P.S.Maria
P.Real
Ronda
Gibraltar Herculeum
Barbaria
Strecho de olim Fretum
Corta
Pilos
S.Lucas
Chipiona
S.Sebaſtiano
Caſtilo S. Philippo
Caſtilo
Puente de Suaco
CALES IN SVLA
Barbareich
Occidentalis Oceani pars
Eſpartel
Al monte
Higuera
Val de Vacas
Palos
Die Armada von Engelandt
Hispania findt newe geſt,
Die Englische Flött thut jhr beſt,
Sampt der Hollendischer Armäd;
Durch brant vnd raub gſchicht groſſer ſchad.
Calis Malis wirdt vbereilt,
Dahin des feiends ſchiff geseilt:
Es gilt zu waſser vnd zu land
Der feiend halt noch vberhand.
Der ausgang diſſer sachen schwer,
Von manchem wirdt erwartet ſehr:
Gott gebe das die grechte sach
Obsige durch ſein Gottliche rach.
Anno domini 1596
Menſe Junio et Julio.

SIEGE OF CADIZ, 1596, BY FRANS HOGENBERG

The English fleet attacks Cadiz, with the Strait of Gibraltar at top right. The threat of another Spanish Armada led, as in 1587, to a pre-emptive strike. The English were lucky as the Spanish fleet was dispersed by storms, while they enjoyed surprise before fighting their way past a combined Spanish force of modern galleons and galleys, supported by the guns of Cadiz. A successful opposed landing was followed by the storming of the city. The Spaniards were able to burn merchantmen sheltering in the inner harbour, but the loss was still immense, although the commanders failed to control their troops and Elizabeth I did not get her share. The English then withdrew as they could not sustain a garrison in Cadiz. ABOVE

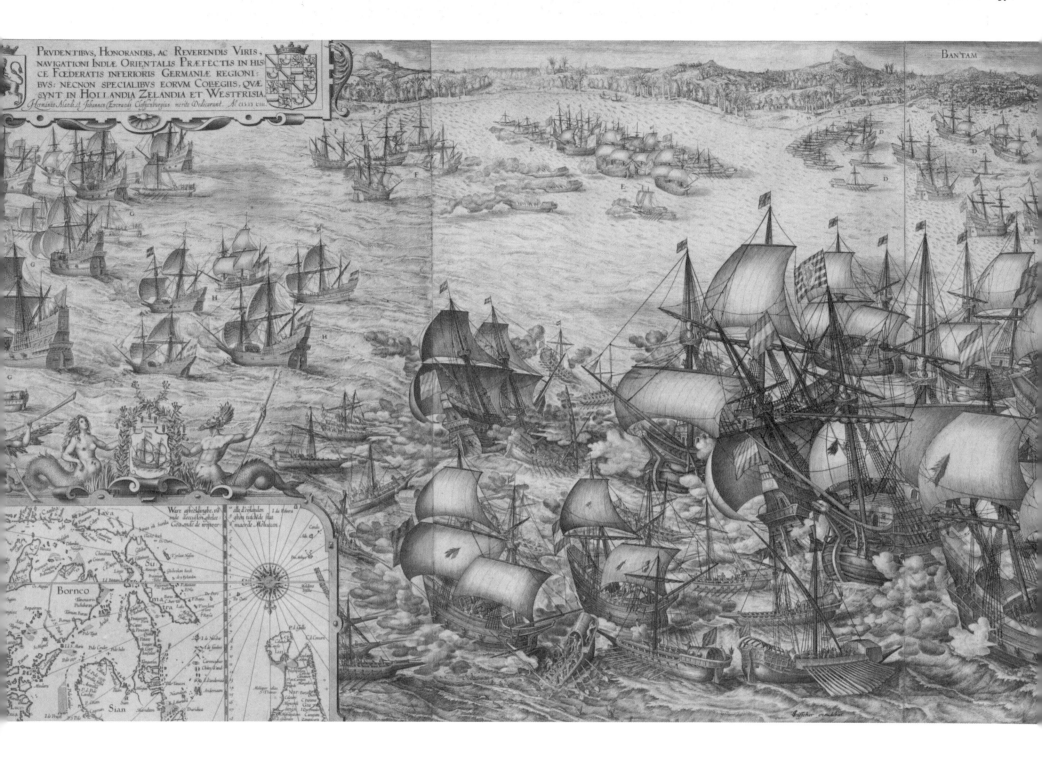

BATTLE FOR TRANSOCEANIC TRADE, 1601

Due to Philip II of Spain's takeover of the Portuguese empire in 1580, the war between the Dutch and Spain came to involve frequent attacks on the Portuguese empire. In December 1601, the Dutch defeated a far larger Portuguese fleet off Bantam in Java, gaining control of the spice trade there. The battle is explained in this illustration in a series of episodes interpreted by means of the use of letters. The Dutch East India Company used force to link trading zones and create monopolies, thus enhancing economic specialisation and exchange. Having defeated the Portuguese, the Dutch defeated the forces of the Sultan of Bantam in 1619 and the town renamed Batavia (now Jayakerta), became the centre of their power. **ABOVE**

BATTLE OF GIBRALTAR, 1607 The Dutch fleet of twenty-six warships under Jacob van Heemskerk surprised the Spanish fleet at anchor, destroying most of it within four hours. Dutch naval power rested in large part on general maritime strength and dynamism. The United Provinces became an entrepôt for European and global networks of trade, in which comparative strength brought the Dutch profit. Economic strength became military power, but the background was a war for survival with Spain. RIGHT

BAYA DE GIBR

arhafftige Abbildung

derbahrlichen / Schifffftreits / zwifchen

ada des Königs von Hifpanien und etzlichen

sfchiffen der Herrn Staden / unterm Com-

do des Admirals Iacob Heimfkirchen

on Amfterdam: gefchehen auf der Re-

de von Gibralter / den 25 Apri-

lis im Jhar 1607.

LTAR

Spaensche
Admirael

Spaensche
schepen

Den Admirael
Heemskerck

Capiteyn Lambert

Casteel

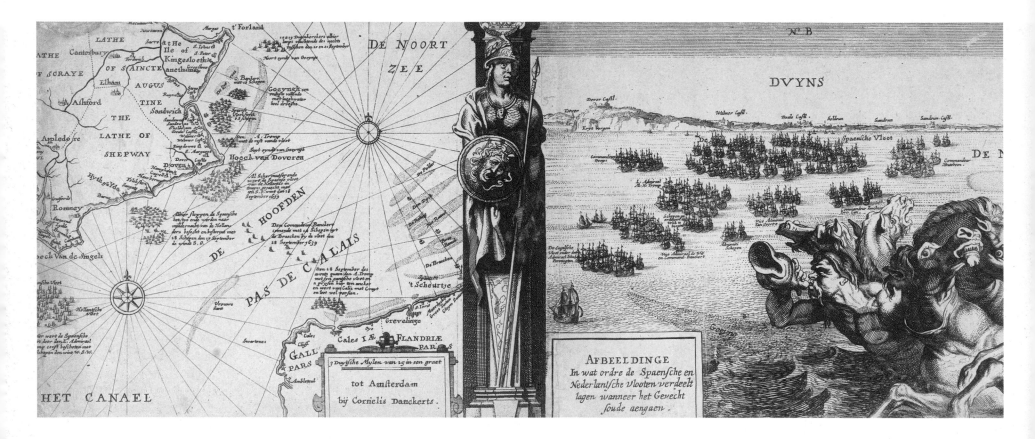

THE BATTLE OF THE DOWNS, 1639, PUBLISHED 1640–45 At war with Spain, the Dutch sought to cut their maritime links with the Spanish Netherlands (Belgium). This battle demonstrated that, for all their improvements in ship design, the Spaniards had still not adopted a policy of naval combat with artillery, and clung to the tactics of boarding, possibly due to the tendency for military commanders to hold authority over their naval counterparts. Having been defeated, the Spanish took refuge in the Downs off the English coast, only to be attacked with heavy losses by Admiral Tromp. In the confined waters, the Spaniards were vulnerable to fireships and to the more manoeuvrable Dutch warships. **ABOVE**

BATTLE OF BERGEN, 1665 As part of the Second Anglo-Dutch War, the English fleet attacked the Dutch fleet in Bergen, Norway. Norway was ruled by Denmark and its fort guns backed the Dutch, preventing an English victory. **RIGHT**

Bergen in Norway

Nor = Way

The Names of his Majes[t] Ships of Warre at B.
a. y^e Norwich ....   h. y^e Garnsey
b. y^e Prudent Mary ...   i. y^e Pembroke
c. y^e Breda ....   K. y^e Revenge
d. y^e Foresight ......   L. y^e Coast Fregat
e. y^e Bendish ....   m. y^e Guinea
f. y^e Happie Retume   n. y^e Greyhound o. y^e Pry
G. ye Saphir ....   p. Sosiette, q ye Golde

Werkeneße.

Hornes

Part of Diesholme

The Chanel from Getts-foed

Discription of ye chief Places about Bergen
1. the Castle   6. a Fort. 34 Guns 1500. Me[n]
2. a platforme of wood 7. ye Custome House
3. a platforme newly made 8. ye Buy,
4. ye seueral Holla[n]ders Guns 9 ye Bay or Port of B.
5. a Smal old Castle   10. ye Towne of Berger[n]
Described from ye Life in Augu[st] 1665 by GH:

DUTCH ATTACK ON CHATHAM, 1667 A sketch of the position of the Dutch fleet on 20–23 June 1667. The Second Anglo-Dutch War (1665–67) saw heavy English naval blows on the Dutch, notably in the battles of Sole Bay (1665) and the Two Days Battle (1666), but the financial strain of the war ensured a lack of preparedness in 1667. This was exploited by the Dutch in a successful attack on the English warships anchored in the River Medway. This meant that the war left a very different political impression to the First Anglo-Dutch War, that of 1652–54. RIGHT

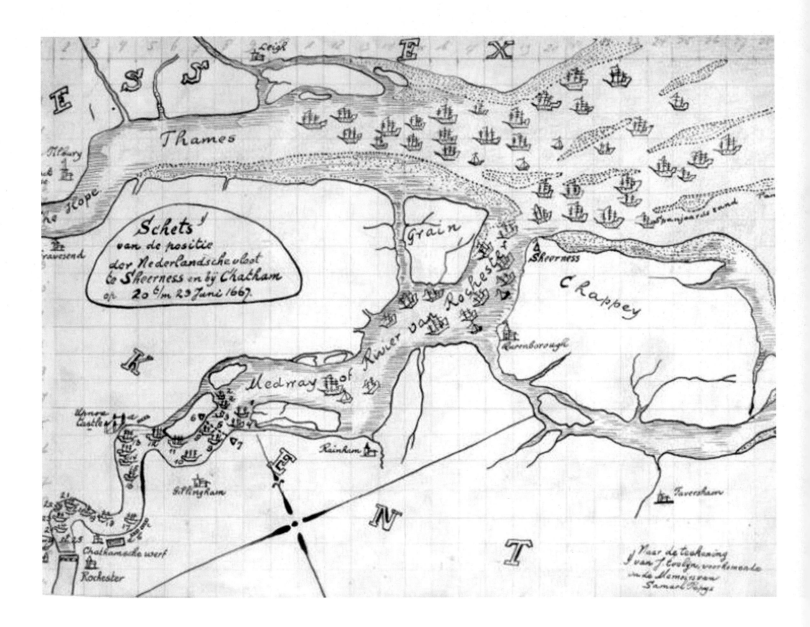

WAR ON TRADE, 1672 A well-sustained action by a small squadron under Sir Robert Holmes against the Dutch Smyrna convoy returning from the Mediterranean launched the Third Anglo-Dutch War (1672–74). The drawing, heavily annotated in a Dutch hand, with details of the first day of the battle, captures the difficulties of retaining position, and the obscuring nature of the black powder used by the guns. ABOVE

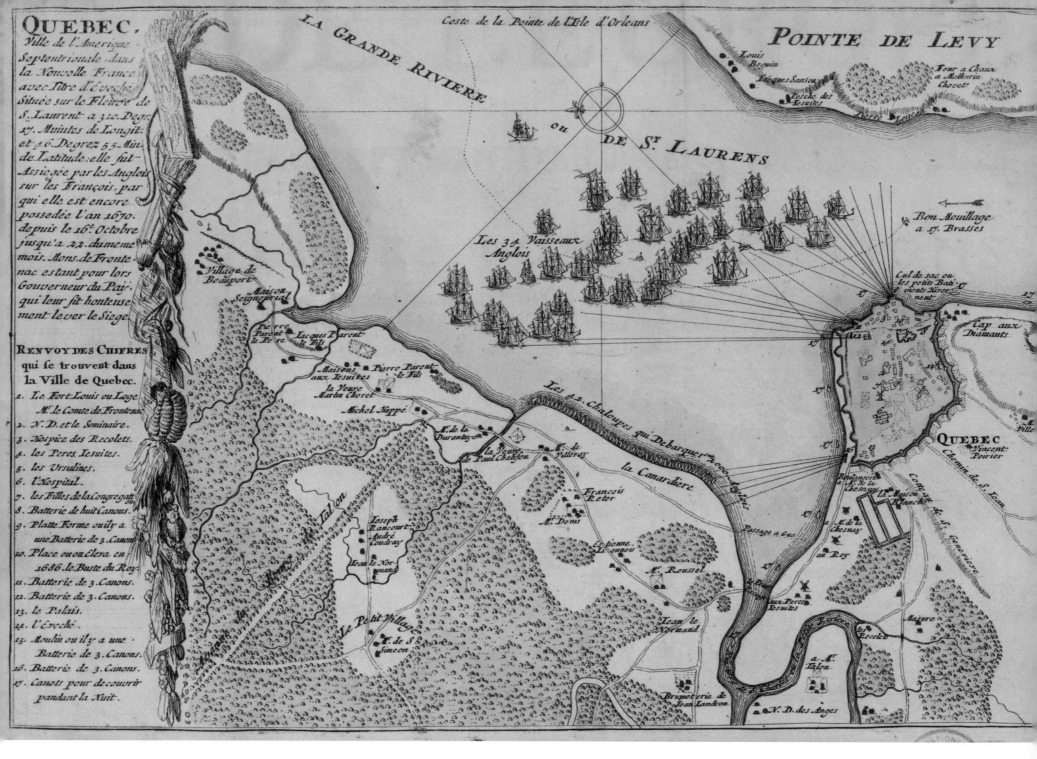

ENGLISH SIEGE OF QUÉBEC, 1693 Naval power projection was crucial to successive English/British attempts to conquer French bases. Québec was established by the French in 1608, only to be starved into surrender by an English force in 1629, before being returned to the French in 1632 after a peace agreement. Larger British forces failed in their attacks in 1693 and 1711, the first in large part due to disease, the second due to a night-time shipwrecking. In 1759, British forces finally captured the city, with James Wolfe's expeditionary force benefiting from a prior surveying of the St Lawrence River by James Cook. **ABOVE**

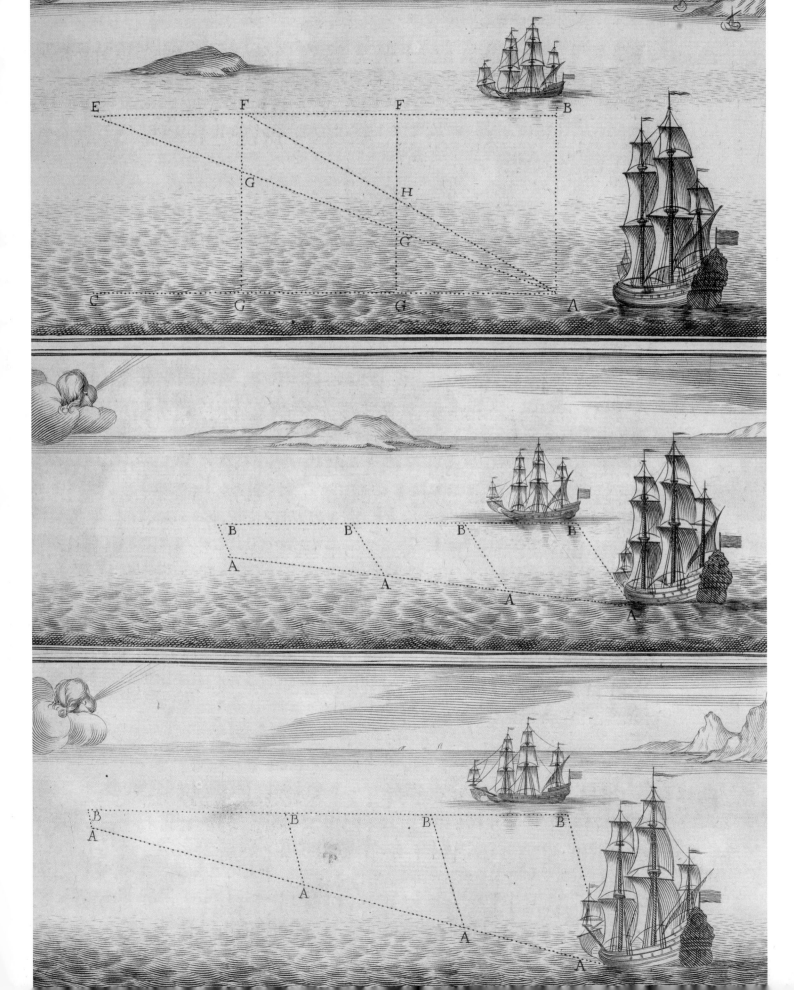

ILLUSTRATION OF NAVAL TACTICS FROM *L'ART DES ARMÉES NAVALES* **BY PAUL HOSTE, 1697**

Paul Hoste's work was an attempt to systematise naval tactics, one that matched the efforts made for land warfare, notably by Vauban. In practice, such geometric formations did not survive contact with the enemy as the French had discovered in their major defeat off La Hougue in 1692. Moreover, Hoste's work was more pertinent for the major French naval build-up from the late 1660s, and did not capture the impact of the economic crisis of 1693–94. With finances under pressure, and influenced by the limited benefits from recent naval operations, the French refocused their naval strategy from the *guerre d'escadre*, the war of squadrons in which they had sought battle, to the *guerre de course*, in which privateering attacks on English and Dutch trade took top priority. **LEFT**

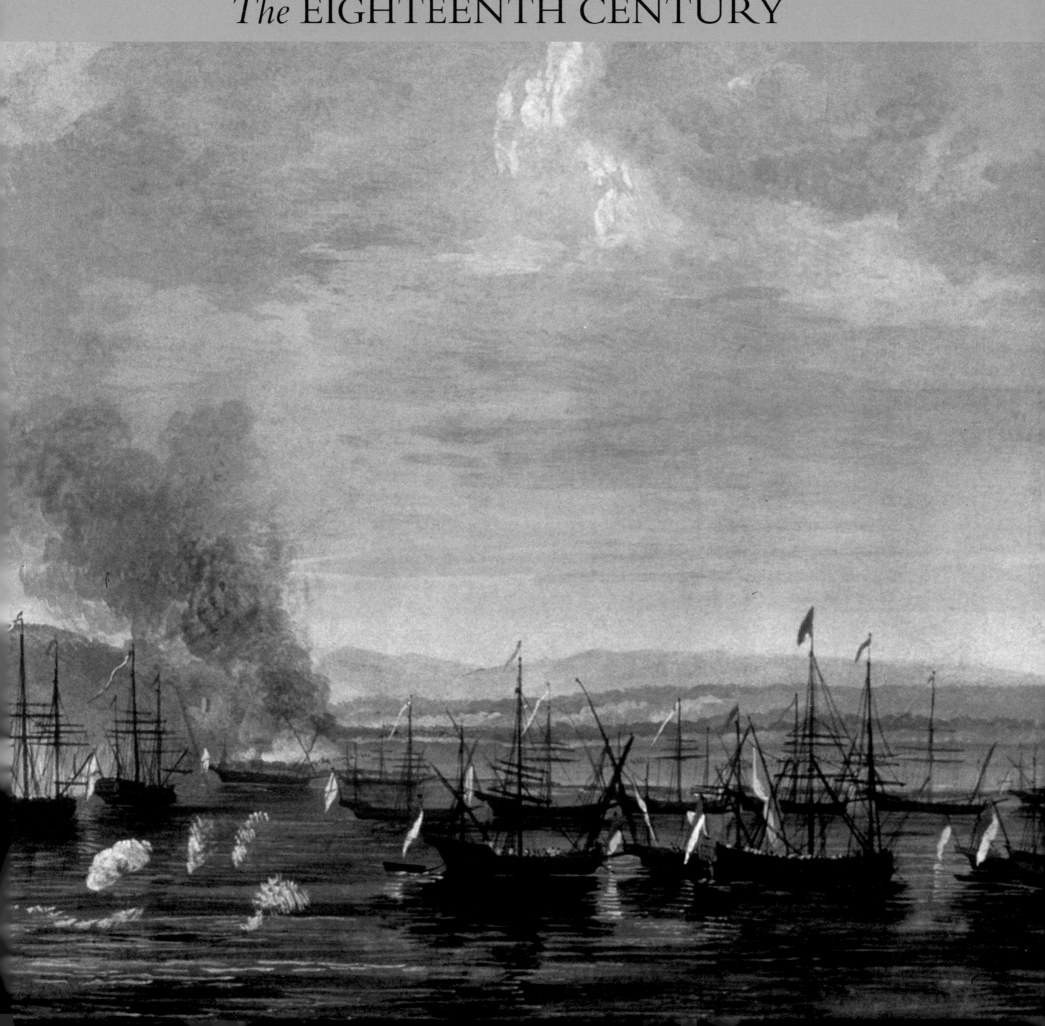

BOMBARDMENT OF COPENHAGEN IN 1700 BY ENGLISH, DUTCH AND SWEDISH FLEETS Coastal cities were highly vulnerable to bombardment from the sea. The Danes, with twenty-nine ships, tried to block the Swedish fleet from joining the twenty-three-strong Anglo-Dutch fleets, but the Swedes managed to slip through on 14 July 1700 by sailing through the shallow part of the Sound at Flintrännan (the eight biggest ships had to be left behind, four after running aground) instead of using the safer main channel of the Sound, and participated in the bombardment of Copenhagen. This was designed to keep the Danish fleet at bay while the Swedish army landed nearby. The Swedes landed safely and Denmark was driven to leave the war. OPPOSITE

A notable book was delivered to the Members of Parliament, with a chart annexed of the Mediterranean Sea, whereby it demonstrably appears of what importance it is to the trade of Great Britain, that Sicily and Sardinia shall be in the hands of a faithful ally, and if possible not one formidable by sea. That these two islands lie like two nets spread to intercept not only the Italian but Turkey and Levant trade … that should the naval power of Spain increase in the manner it has lately done, that kingdom may assume to herself the trade of the Mediterranean Sea, and impose what toll she pleases as the King of Denmark does at Elsinore.

Worcester Post-Man, 21 November 1718

In 1718, a map appeared in a key role in an explanation of naval policy. The British government was keen to win domestic support for the dispatch of a fleet to the Mediterranean in response to the Spanish invasion of Sardinia in 1717 and in Sicily in 1718. The publication of a pamphlet by Reeve Williams, *Letter from a merchant to a Member of Parliament, relating to the danger Great Britain is in of losing her trade, by the great increase of the naval power of Spain. With a chart of the Mediterranean Sea annex'd,* was actively aided by the government in the shape of financial support from the Lord Chancellor, the 1st Earl of Macclesfield. In the pamphlet, which prominently included the map, Williams explained the need for information:

The late action between His Majesty's fleet and that of Spain [the Battle of Cape Passaro off Sicily which Britain won] is become the entertainment of most conversations; but the misfortune is that there is not one in five hundred of the persons who thus entertain themselves that has any just idea of the good consequences thereof. For though the words Sicily

and Sardinia are often in their mouths, they scarce know in what part of the world those islands are situated.

It was not necessary to have the map in order to understand the argument about naval power, but the map, which appeared at pride of place in the pamphlet, helped both to clarify the argument and to lend it weight. This example suggests that maps about naval power thus became a component not only of public entertainment, but also of public education, the potential of which affected the context within which policy could be discussed in states with a defined and articulate political nation. It is also, however, necessary to note that the overwhelming majority of British (and other) pamphlets on foreign affairs did not include maps. Moreover, very few newspapers ever included a map. There was no equivalent in the public to the teaching of navigation to maritime and naval officers. It was still complicated to print a map and expensive for newspapers to do so.

Government backing for the publication of this pamphlet was an aspect of the overlap of public and private, government and entrepreneurialism, that was so important to the map world in Britain, as earlier in the Low Countries. In these states, map-makers depended heavily on the private sector for sales and finance. This ensured that the views of customers were crucial, or at least those views as understood by entrepreneurial map-makers pursuing competitive advantage. Commercial pressures in Britain led to a consolidation of map publishing, and consumers were left to be their own judges of quality. At the same time, commercial map publishers could draw on government sources. Thus, Thomas Jefferys, Geographer to the King, was granted privileged access to, and allowed to print, manuscripts held by official map-making agencies such as the Admiralty and the

Das
Von denn dreyen Alliirten,
als
Englisch-Schwedisch-u.Holländischen
Flotten
Bombardirte,
doch nicht
Beschädigte/
COPENHAGEN
Anno 1700.
den 20.6.u.26.9.Julii.

Erklærung der Zahlen.

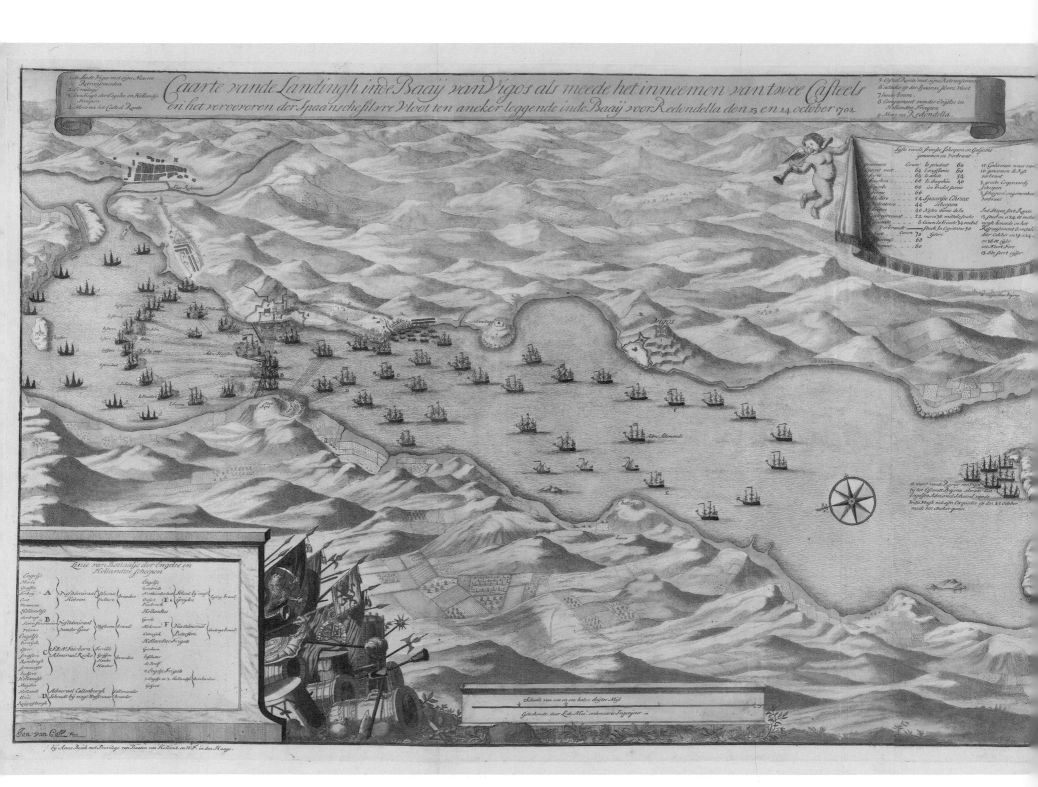

Board of Trade, the latter the key body for the administration of the colonies. Institutional support made it possible to begin with a kind of standard in map design and scale.

Moreover, ministers, officials, politicians and other prominent figures owned maps. An instance of the ownership of maritime maps is shown by the copy of the *Atlas maritimus and commercials; or, a general view of the world* by Nathaniel Cutler and Edmond Halley that was published in London in 1728 and belonged to John Clevland. A Commissioner of the Royal Navy from 1743 to 1746, Clevland became joint secretary to the Admiralty from 1746 and sole secretary from 1751 until his death in 1763. This work included a coasting pilot for mariners with fine charts of most trading areas of the world. Governments, indeed, developed important collections of maps. The French Ministère de la Marine (Ministry of the Navy) established its Dépôt des Cartes et Plans (Office of Maps and Plans) in 1720 (well before most states), but did not use its power as an official agency to implement a kind of standardisation of maps.

CHALLENGES IN MAP-MAKING

The production of accurate maps was necessary because most of the world's waters were uncharted. In *The construction of maps and globes*, John Green observed of Europe in 1717: 'Tis observable, that not only the sea coasts, in two several maps of the same parts commonly differ strangely from each other; but also rarely ever any agree in that respect with the sea charts.' There was also a widespread lack of standardisation in measurements, notably a standard of length. This helped make it complicated to compare maps of the same part of the world. This situation caused major problems for blockading ships, and the blockade of French ports was a key role for the British

navy. It was necessary in order to protect Britain from invasion, such as a recurrence of the sort of invasion attempted in 1588 and experienced in 1688.

These problems were accentuated by difficulties in establishing location at sea and the related problem of timekeeping. The difficulty of establishing a location in part rested on a lack of knowledge of the exact time and thus of a reliable measurement of travelled distance. For example, in 1708, when a French squadron carrying the so-called James VIII and III, the Jacobite Pretender to the crowns of Scotland and England, succeeded in avoiding the British blockading squadron off Dunkirk in the mist, reaching Scottish waters before its pursuers, the initial landfall was made not at the mouth of the Firth of Forth, but, as a result of error, 100 miles further north. As a consequence, the French lost the initiative, British pursuing warships came up and the French returned to Dunkirk, their only port on the North Sea. James was not landed and the projected invasion did not occur. Nevertheless, the plan underlined the degree to which the price of liberty was eternal vigilance, at least as measured by the availability of a strong and well-directed navy.

In another instance illustrating the difficulties of fixing position, George Anson of the Royal Navy nearly ran aground on the island of Tierra del Fuego. Dead reckoning had put his position more than 300 miles out to sea. He pressed on to enter the Pacific, seize a Spanish treasure ship in 1743 and sail round the world, a triumph of navigation that helped provide the heroic tales important to Britain's national self-image as a naval power.

As a result of the difficulties in mounting blockades, the British navy made major efforts to produce accurate charts of the waters off France and Spain, although blockading ships still ran aground. This owed something to the difficulties of keeping away from the

VIGO, 1702, BY JAN VAN CALL In 1702, an Anglo-Dutch fleet under Admiral George Rooke failed to intercept the Spanish New World treasure fleet at sea, but attacked it at Vigo. The treasure fleet was protected by a boom and strong batteries, but the southern battery was captured by an amphibious force and the boom was broken by HMS *Torbay* under heavy fire. The subsequent engagement was decisively won by Rooke's fleet and the French and Spaniards set fire to their ships. English naval strength helped lead Portugal to abandon its French alliance in 1703. **OPPOSITE**

coast when being driven towards it by the wind, a situation that did not change until the onset of steam power in the nineteenth century. During the French Revolutionary and Napoleonic Wars (1793–1815 for Britain), the Royal Navy lost more ships as a result of running aground than due to enemy action. Navies as a rule lost ships to navigational hazards unless they were confined to port.

More problems were encountered in mounting operations far from home waters. When, in 1791, the British government came close to war with Russia in the Ochakov crisis, it discovered that it had no charts for the Black Sea and had to turn for information to its Dutch ally who, characteristically, had more information. Indeed, the lack of knowledge left the British unclear whether the fortress of Ochakov, the crucial issue in the negotiations, really controlled the entrance to the River Dnieper, as was claimed in order to demonstrate its strategic importance. Ochakov is, in fact, on the northern shore of the Dneprovskiy Liman, a nearly landlocked section of the Black Sea into which the estuaries of both the Bug and the Dnieper open. Ochakov is situated at the narrow strait which forms the seaward entrance of this section or bay, but the British lacked adequate maps and coastal charts to show this. These were not the sole deficiencies the British faced. In 1805, HMS *Victory*, Nelson's flagship, had to rely on a French chart that was 40 years old in order to navigate in the western Mediterranean.

CHARTING COASTAL WATERS

Where they could, the British mapped coastal waters. Inshore waters were of greater danger for warships and merchantmen. For example, as far as transatlanticSpanish voyages in the sixteenth or seventeenth centuries were concerned, more ships were lost in entering or leaving the harbours of Cadiz,

Havana and Vera Cruz than crossing the Atlantic.

The sources of initiative for mapping varied, including in the case of Britain the Royal Navy, colonial governments and private individuals. The net effect was one of major improvement. For example, successive surveys of the Carolina coast led to improved maps, especially the coverage of the coast in Edward Moseley's detailed map of 1733. Most notably, having conquered New France (the French-ruled parts of modern Canada) from France in 1757–60, the British surveyed Canadian Atlantic waters. The most prominent surveyor, the Swiss-born Joseph Frédéric Wallet Des Barres, who had been trained at the Royal Military College, Woolwich, was ordered by the Admiralty to survey the coasts of Nova Scotia and Cape Breton Island, which, having been conquered, had been acquired from France under the Treaty of Utrecht (1713) and the Peace of Paris (1763) respectively. In 1777, the first edition of an atlas of Des Barres' navigational charts, the *Atlantic Neptune*, was published for sale. This was an impressive attempt to provide a systematic charting of North American waters, and ultimately comprised 115 charts and maps.

In 1764–81, George Gauld charted the waters of the Gulf of Mexico in response to instructions from the Admiralty which wished to consolidate the recent acquisition of Florida from Spain under the Peace of Paris of 1763. At that stage Florida extended on the Gulf coast as far as the Mississippi. It was necessary to understand all the inlets on the coasts of what are now Florida, Alabama, Mississippi and eastern Louisiana. This charting was also important for Caribbean trade and for an appreciation of the interplay of currents, tides, winds, islands and navigable routes.

The British were also active in Indian waters. In the early 1760s, the highly talented James Rennell charted the Palk Strait and Pamban Channel between India

Veüe du d'Estroit de Gibraltar, et des Enuirons, auec les tranchées du Siege mis en 1704.

Montagnes de Totteguin au Royaume de fez

Camp des mores

Nuestra Señora de Africa

Ceuta a l'Espagne

Sebta

Destroit de Gibraltar

Sud

Ouest

Est

Nort

12 brasses

Tour del hacha.

Nuestra Señora de Europa.

antien Gibraltar

batterie Superieure

la pate

Chasteau

Vieux Mole

Nouueau mole

5 brasses

Nouueau Gibraltar

golfe de Gibraltar

10 brasses

château du Comte Julien que les mores en Espagne.

Algezires

8 brasses

les tranchees

7 brasses

6 brasses

camp des Assiegeans

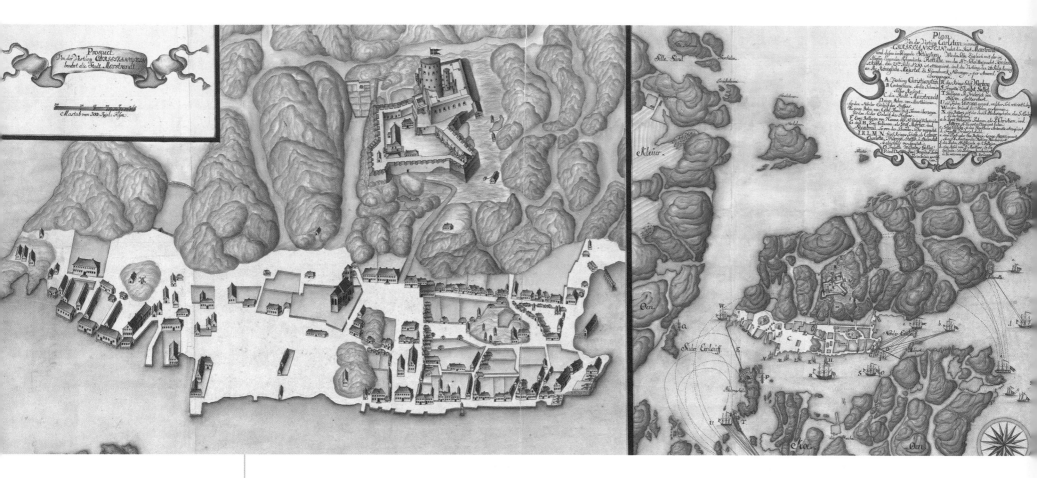

and Sri Lanka (Ceylon), the shoals of which were a major challenge to ships seeking to sail round India, without having to sail round Sri Lanka. Alexander Dalrymple, official hydrographer to the East India Company in 1779, also became hydrographer to the Admiralty in 1795. The Order in Council instructed Dalrymple 'to take charge of such plans and charts as are now or may hereafter be deposited in this office belonging to the Public, and to be charged with the duty of selecting and compiling all the existing information as may appear to be requisite for the purpose of improving the Navigation, and for the guidance and direction of the commanders of Your Majesty's ships'.

CLASSIC PERIOD IN NAVAL WARFARE

The eighteenth century is a classic period of naval history. Alongside dramatic battles that sometimes had only a limited strategic impact, there is the major role of naval warfare in the War of American Independence, especially the Battles of Ushant (1778), the Virginia Capes (1781) and the Saintes (1782), and the extent to which the naval history of the period was discussed by subsequent naval theorists, notably Alfred Thayer Mahan and Julian Corbett. Moreover, there were more naval battles than in the nineteenth century, albeit with the battles not involving most Asian powers, but being very much between Western powers and also Turkey. Contemporary maps of battles themselves were of limited value to understanding what had occurred. They were a static presentation of a more complex set of events, as, for example, with the maps of the Anglo-Bourbon Battles of Malaga (1704), Toulon (1744) and Lagos (1759, fought off Portugal); battles that played an important role in ensuring British naval dominance in the western Mediterranean. However, as with the situation in the sixteenth and seventeenth century, many maps were designed to provide, like pictures, illustrative interest. That the maps published for the public of naval battles were retrospective was crucial in this respect. There was no respect in which they could be useful for new engagements.

The in-time mapping of soon-to-be or, even, current operations that became important in the twentieth century was not done in the eighteenth. There were no facilities for on-board printing of maps, let alone any electronic finding aids. Moreover, reconnaissance of opposing forces was necessarily limited to eyesight, telescopes and visual recognition of ships and flags. The closest to over-the-horizon reconnaissance was provided by observers in the rigging and the masthead. Without radio communication, human intelligence from within enemy harbours was of limited in-time value. Given also the difficulties of forcing battle on opponents who did not want to fight, it was unsurprising that planning was very much *ad hoc*. Nevertheless, with the training of a fleet as a unit, rather than as single ships, the need for planning increased.

However, once the cannon had opened fire, the details of a battle generally became obscure. This reflected the extent to which cannon were using black powder which produced copious quantities of smoke. The lack of standardised timepieces contributed to the difficulties of recording and recovering detail.

Battles frequently arose when the attempts by one side to avoid battle by, for example, sailing into supposedly safe waters, failed. Prime examples were the British victories over the French at Lagos and Quiberon Bay in 1759, respectively in neutral waters on the coast of Portugal and in a rock-strewn bay on the coast of Brittany. These victories, which owed much to skilled command and well-tried and experienced ships, captains and crew, stopped the French attempt to invade Britain. Similarly, Trafalgar in 1805 meant that, having lost so many ships, the French no longer had the capability or the will to mount such an invasion.

NAVAL CONFLICT OTHER THAN BATTLE

As with other periods, the emphasis on battle can lead to a failure to devote due attention to the many forms of naval conflict, ranging from blockades to commerce raiding and its corollary, commerce protection. Amsterdam Marine insurance rates in the 1760s to 1770s indicate that war was seen as the major threat to trade, with rates far higher in wartime.

Moreover, amphibious operations remained important, and increasingly so as imperial possessions became more significant in power politics. The key demonstration of British naval power and long-range naval capability occurred in 1762 when expeditions captured Havana and Manila from Spain. Each expedition was dependent on naval strength. Captain John Elphinstone of HMS *Richmond*, who led the fleet through difficult waters to Havana, presented the Admiralty with a chart of the north side of Cuba that showed the route of the fleet and the degree of detailed information the Royal Navy could command. Naval strength and accurate information were also crucial to the capture of Louisbourg (1758) and Québec (1759) from France. Conversely, the reduction of British naval

MAPS OF THE GREAT NORTHERN WAR (1700–21) SHOWING DANISH NAVAL ATTACK ON A SWEDISH POSITION The maps capture the different scales at which operations have to be considered. In order to defend Marstrand Island from an expected Danish attack, most of the Swedish Gothenburg squadron was based there: Marstrand is north of Gothenburg. In July 1719, the Danes attacked and bombardment from land and sea played a significant role in the campaigning. The Danes pressed on to land on Marstrand and capture the town, which led to the scuttling of the Swedish warships. The Swedes retreated into the fortress of Karlsten but the artillery officers refused the commandant's instructions to bombard the town as they came from it. The Swedes then surrendered on the promise of preserving their freedom. The commandant was court-martialled and executed when he returned to Sweden. OPPOSITE

DEFENCE OF TOULON, 1707 This was a highly
successful combined operation against
Toulon with the total elimination of
France's Mediterranean fleet thanks to
an Anglo-Dutch naval bombardment
which was combined with a siege by
Austrian and Piedmontese forces. The
siege was stopped when it appeared clear
that the city would not fall speedily and,
instead, could resist until the arrival of
overwhelming French forces. During the
siege, the Anglo-Dutch fleet played a key
role in supporting the siege, providing
cannon, supplies and medical care.
The Toulon campaign indicated both
the growing importance of amphibious
operations and the extent to which the
key issue was not the seizure of territory,
but the achievement of particular
strategic goals in the shape of
destroying the fleet. OPPOSITE

capability in North American waters in 1781 was crucial to British failure in the War of American Independence. The Royal Navy was unable, in the Battle of the Virginia Capes, to break the French blockade and relieve the British force at Yorktown, which led to the surrender of this force to its American and French besiegers.

STATIC CAPACITY

In terms of both power projection and conflict, the British capability to move warships and large numbers of troops as far as the Americas and India was important, but there was no significant change in naval capability in this period, and certainly nothing to match the changes over the following century. In terms of propulsion, structure, armament, manpower, supply and organisation, there was fundamental consistency. Individual changes could improve effectiveness (for example, the copper-bottoming of ships to resist the inroads of marine worms on wooden timbers, a particularly serious problem in tropical waters) and these changes could be important. However, they were less significant than the ability to maintain supply systems and the skill of individual commanders. The British proved adroit with both. By 1762, the Royal Navy had about 300 ships and 84,000 men, a size that reflected political support, the growth of the mercantile marine, population, economy and public finances, as well as a heavy shipbuilding programme during the Seven Years' War. The resources required were immense. Launched in 1764, HMS *Triumph*, a standard 74-gunner, needed 3,028 loads of timber, each fifty cubic feet. The 100-gun HMS *Royal George* needed 5,760.

NAVAL FORCES AROUND THE WORLD

A maritime populism became even more significant than hitherto to Britain's self-image. Accordingly, the idea of Britain ruling the waves was proclaimed with great zeal. This idea helped drive out the alternative image of success on land, although, in practice, the army enjoyed much success, notably under the Duke of Marlborough in the 1700s and, again, in 1759 at Minden and Québec. However, armies were associated with autocratic traditions and tendencies, such as Oliver Cromwell and Hanoverianism in Britain, and this encouraged a stress on the navy that was unmatched in the world, a stress only matched by the (weaker) Dutch.

This stress was not to be taken on board in the new United States after it gained independence from Britain in 1775–83. There were American politicians who thought it important to develop a navy, notably Alexander Hamilton who regarded this as the way to protect American trade. However, there was also a potent hostility to this approach. A navy was seen by politicians such as Thomas Jefferson, the third president, as a means toward a powerful centralised state on the British model, one that they presented as alien to American liberty and nationhood. Instead, the emphasis was on local self-reliance in the shape of the militia. Jefferson favoured a naval equivalent in the form of gunboats, as opposed to the ships of the line supported by his rival Hamilton. Ultimately, the challenge from European naval power, especially from the British during the War of 1812–15, encouraged the development of an American navy, but, with the exception of the major build-up during the American Civil War (1861–65), it remained relatively small and weak until the late nineteenth century.

Elsewhere, naval forces remained small. Those of China and Japan were essentially for coastal defence against piracy. Turkey had a significant navy, but it was defeated, most prominently by the Russians at Cesmé

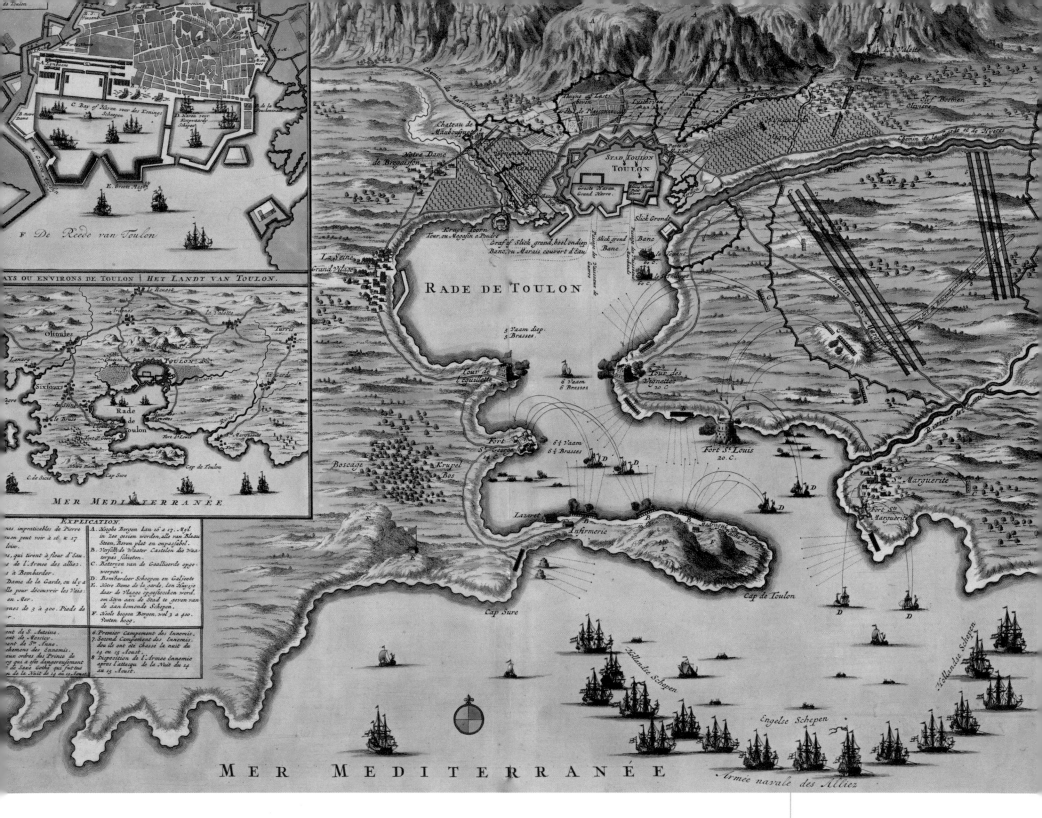

off the island of Chios in 1770, a victory in which the use of fireships played a major role, and again in the late 1780s. A number of Indian rulers, notably Tipu Sultan of Mysore, attempted to develop naval power, but were thwarted by the strength of Britain and by their own focus on land strength. No African ruler had significant naval power, and that of Pacific rulers was only local.

Britain demonstrated the broadly-based foundations of naval power. It was a function of ships, manpower, bases, logistical support, funding and political support, each of which was intimately related to the others. Moreover, there was a need for consistency, a consistency that in particular depended on political support in the face of the generally more potent demands of armies. Again, Britain proved distinctive in this fashion.

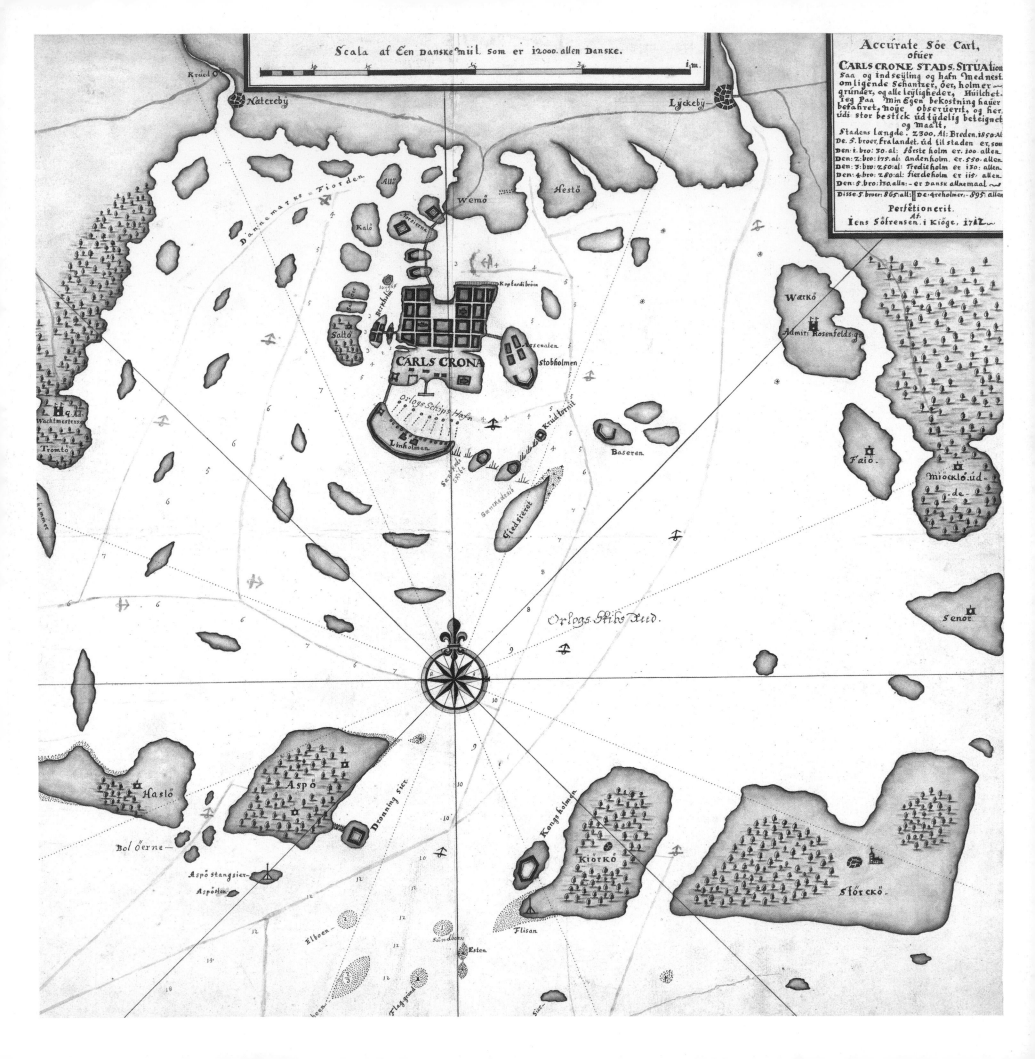

Scala af Een Danske miil, som er 12000. allen Danske.

Accurate Söe Cart,
ofuer
CARLS CRONE STADS. SITUAtion
saa og indseyling og hafn med nest
omligende Schantzer, öer, holmer
grunder, og alle leyligheder, Huilchet
ieg paa min Egen bekostning hauer
befahret, noye obseruert, og her
udi stor bestick udtydelig beteignet
og maalt,
Stadens længde. 2300. Al: Breden 1850 Al
De. 5. broer, Fralandet ud til staden er som
DEN 1. bro: 30. al: försте holm er 100. allen
DEN 2. bro: 175 al: Andenholm. er 550. allen
DEN 3. bro: 250 al: Tredieholm er 130. allen
DEN 4. bro: 280 al: fierdeholm er 115. allen
DEN 5. bro: 130 alln: er dansk allnemaal
Disse 5. broer 865 all. De 4 reholmer 895 allen
Perfetionerit.
Af
Iens Söfrensen. i Kiöge. 1712

Krued O.
Næterbÿ
Lÿckebÿ

Dannemarks = Fiorden
Allö
Kalö
Wemö
Hestö
Nereierna
Berholm
Koplardi broen
Saltö
Warkö
Admir: Rosenfelds g.

CARLS CRONA
Arsenalen
Stobholmen

Hq. Ad. Wachtmestersg.
Tromtö
Orlogs Schips Hafn
Krudtornit
Limholmen
Baseren
Faiö
Miocklö ud de.

Sandbet Skib
Santkedesis
Giedsieret

Senor

Orlogs Stiks Aud.

Haslö
Bol öerne
Aspö
Dronning sier
Kongsholmen
Störckö
Aspö stangsier
Aspösten
Kiörkö
Elboen
Sundboen
Flisan
Esten
Flaggrund

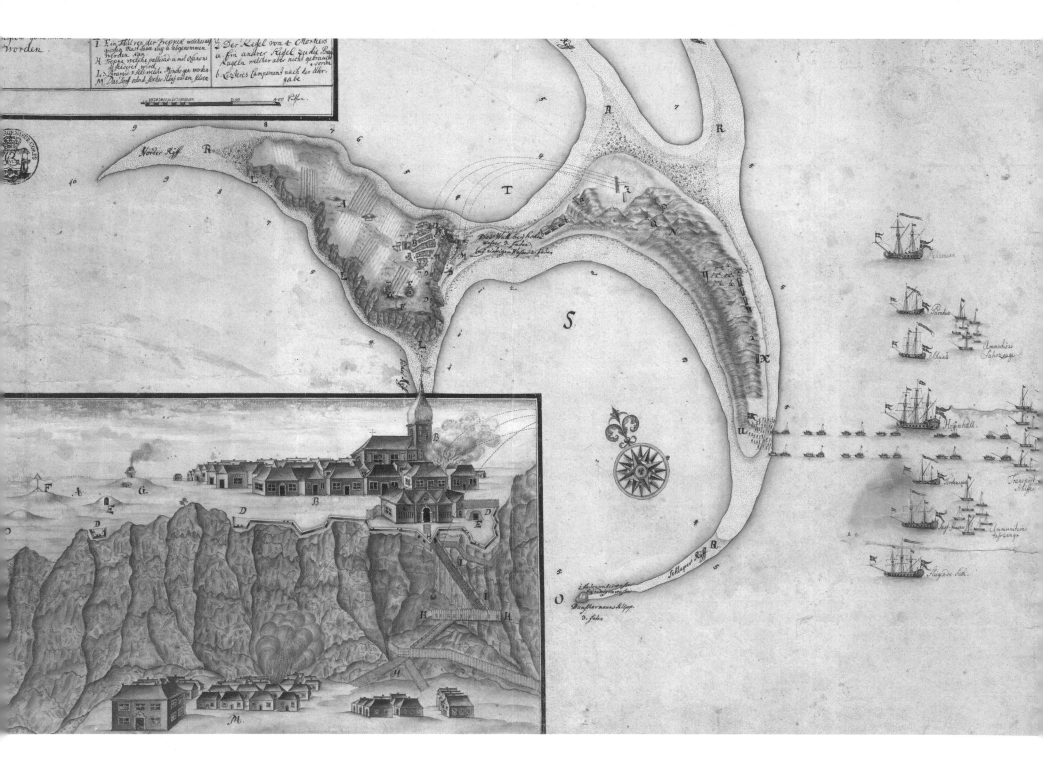

THE SWEDISH NAVAL BASE OF KARLSKRONA, 1712
This map was made during the Great
Northern War (1700–21). Its author was
Admiral Sørensen, who later headed up
the naval map archives. The fortress port
was defended by offshore island bases.
LEFT

**SUCCESSFUL DANISH AMPHIBIOUS ASSAULT ON
HELIGOLAND DURING THE GREAT NORTHERN
WAR, 1714** Heligoland was a part of the
Duchy of Holstein-Gottorp, an ally of
Sweden but as a result of the war,

Denmark acquired it. It was captured by
Britain in 1807 and then exchanged with
Germany for Zanzibar in 1890. Map and
picture combine to indicate the impact
of bombardment. **ABOVE**

PLAN DER BELAGERU[NG] VON FRIDRICHSHAL[L]

angefangen von CARL dem XII glorwürd. andenckens der Schweden Go[tt] Wenden König mit 10000 Mann 36 Can: u. 18 Mort. d. 11 Nov. aufgehoben d. 1[...]

Danische Postirung so sich retiriret

Königl. Schwedisch Lager so zu Ende des Novembers bezogen und den 20 December Anno 1718 wieder aufgehoben worden

TISTEDALENS ELBE

THEIL DES KÖNIGR EICHS NORW[EGEN]

Schwedische Batterien von 18 Canonen

Christiania

Fridrichstat

Holtzernes Haus worin der König von Schweden nach Eroberung der Schantz Guldenlow sich aufgehalten

Das Fort Guldenlow erobert d. [...]

Kleine Brand Berg

Fels Oglan

Kaggi

Schwedische Batterie die oben nach die Stadt Fridrichstein kommen seind

Laufende der hinter Fridrichstein König C. H. Dec Nacht um 11 Uhr durch einen Cartätsch Schuss das Leben verlohren

Ein gepflasterter Weg Kliwen genant

Fels Pa[...]

Königl. Danische Postirung so sich aber zuruck gezogen

Klanne

Nort Seite

Materialhof

Fridrichstein

Star Tornet

Ober Berg

Röo

Ros brücken

Die offene Stadt und Hafen FRIDRICHSHALL

Knardahl

Diser weg is[...]

IDEFIORD

In dieser Gegend ist die Schwedische und Danische Flottille offt aneinander gewesen

Schwedische Postirung

Schwedische Postirung

Wen[...]

Savöe

Bartve Insula

Schwedische Postirung

Ecksvrg

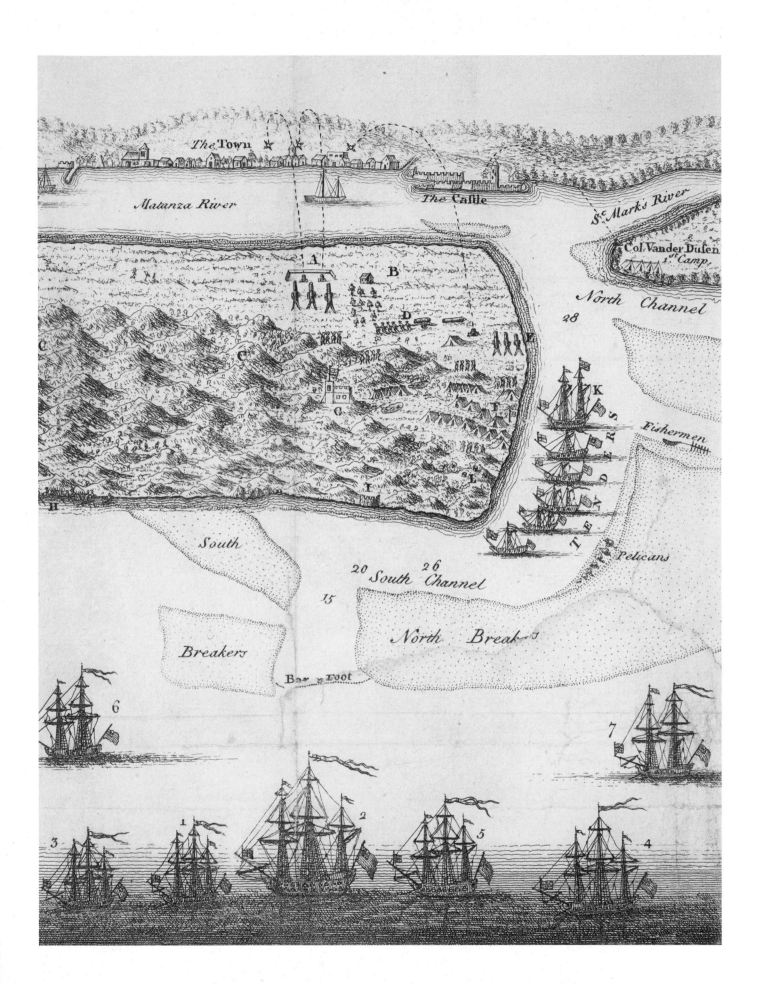

The Town

Matanza River

The Castle

St Marks River

Col. Vander Dusen
1st Camp.

A

B

D

C

C

North Channel

28

G

K

T
L

H

T

Fishermen

L

T E N D E R S

Pelicans

South

20
South
26
Channel

15

Breakers

North Break

Bar Foot

6

7

3

1

2

5

4

THE SWEDISH SIEGE OF DANISH-HELD FRIDRICHSHALL IN NORWAY, 1718 This was the siege in which Charles XII of Sweden was killed. It was accompanied by an inshore naval action. OPPOSITE

ATTACK ON FLORIDA, 1740, FROM *GENTLEMAN'S MAGAZINE* The readers of the *Gentleman's Magazine* were given a clear view of English naval superiority off St Augustine, but the attack failed. James Oglethorpe, the governor of Georgia, planned a methodical siege, but the South Carolina Assembly insisted on a short campaign. The naval blockade failed to prevent the arrival of supply ships, the well-fortified and ably-defended Spanish position resisted bombardment and, in the face of desertion and disease, Oglethorpe retreated. In 1702 St Augustine had also successfully resisted British attack, being relieved by a fleet from the Spanish colony of Cuba. LEFT

ATTACK ON GHERIA, 1756 On 12 February 1756, a British naval squadron under Rear-Admiral Charles Watson demanded the surrender of Gheria, a stronghold on the west coast of India of the Angrias, a Maratha family whose fleet was a factor in local politics and had been used for privateering attacks on European merchantmen. When the Indians opened fire, Watson 'began such a fire upon them, as I believe they never before saw, and soon silenced their batteries, and the fire from their grabs [ships]'. The five-hour bombardment also led to the destruction of Tulaji Angria's fleet, which was set ablaze with shells. The next day, the British warships closed in to bombard the fort at pistol-shot distance in order to make a breach in the wall for storming and this breach swiftly led to its surrender. Watson noted that 'the hulls, masts and rigging of the [British] ships are so little damaged, that if there was a necessity we should be able to proceed to sea in twenty-four hours'. Attacks in 1718 and 1720 had failed. In 1756, Watson co-operated with Robert Clive and with Maratha troops. **RIGHT**

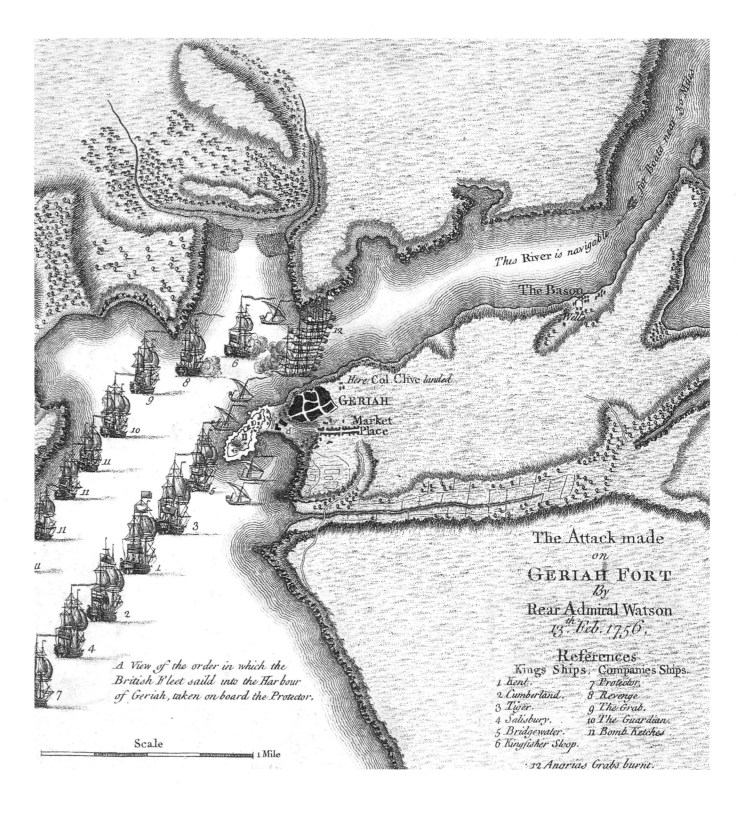

This River is navigable

The Bason

Here Col. Clive landed

GERIAH

Market Place

The Attack made on GERIAH FORT By Rear Admiral Watson 13th Feb. 1756.

A View of the order in which the British Fleet saild into the Harbour of Geriah, taken on board the Protector.

References
Kings Ships.   Companes Ships.
1. Kent.   7. Protector.
2. Cumberland.   8. Revenge.
3. Tiger.   9. The Grab.
4. Salisbury.   10. The Guardian.
5. Bridgewater. 11. Bomb Ketches.
6. Kingfisher Sloop.

12 Angrias Grabs burnt.

Scale   1 Mile

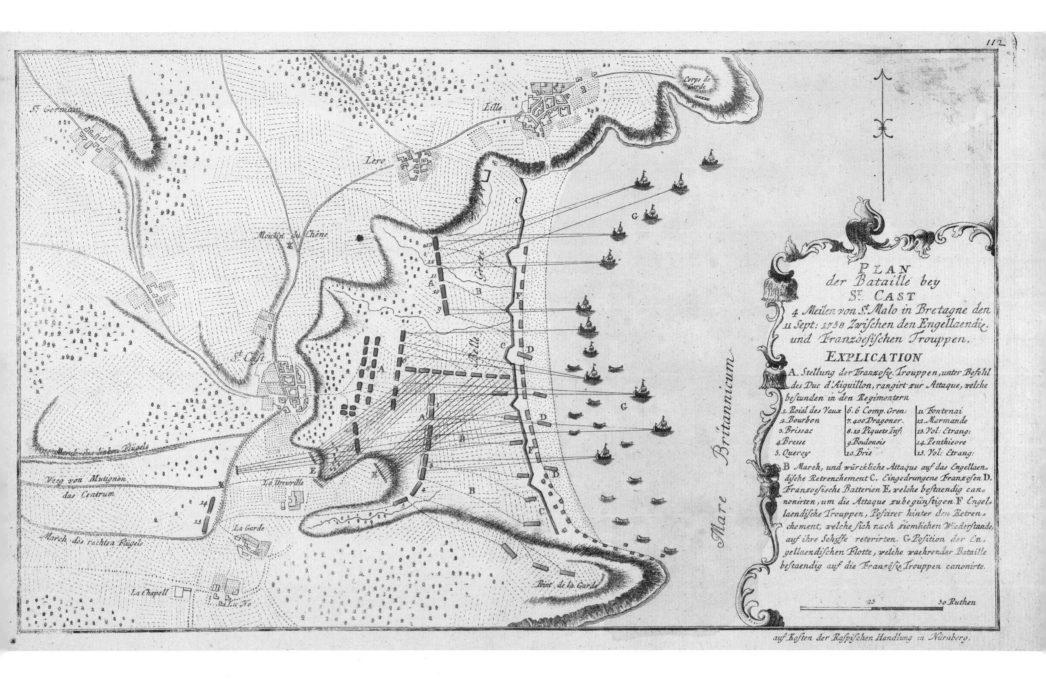

112

PLAN der Bataille bey St. CAST
4 Meilen von St. Malo in Bretagne den 11 Sept: 1758 Zwischen den Engellaendie. und Franzoesischen Trouppen.

EXPLICATION

A. Stellung der Franzosie. Trouppen, unter Befehl des Duc d'Aiguillon, rangirt zur Attaque, welche bestunden in den Regimentern

1. Roial des Vaux	6. 6 Comp. Gren:	11 Fontenai
2. Bourbon	7. 400 Dragoner.	12. Marmande
3. Brissac	8. 12 Piquets Inf:	13. Vol: Étrang:
4. Bresse	9. Boulonois	14. Penthieore
5. Quercy	10. Brie	15. Vol: Étrang:

B March, und würckliche Attaque auf das Engellaen. dische Retrenchement C. eingedrungene Franzosen D. Franzoesische Batterien E welche bestaendig can. nonirten, um die Attaque zu begünstigen F Engel. laendische Trouppen, Posiret hinter den Betren. chement, welche sich nach ziemlichen Wiederstande, auf ihre Schiffe reterirten. G Position der En. gellaendischen Flotte, welche waehrend der Bataille bestaendig auf die Franzosie. Trouppen canonirte.

25   50 Ruthen

auf Kosten der Raspischen Handlung in Nürnberg.

AMPHIBIOUS ASSAULT IN BRITTANY, 1758 A German map, published in Nuremberg, a major map-making centre, of the British coastal assault on St Cast in Brittany in September 1758. Designed to attack the French privateering base at St Malo, the British had to re-embark with the loss of 750 men in the face of a larger French army. The previous month, there had been a successful attack on the port of Cherbourg and its fortifications had been destroyed. These attacks were mounted in part to take pressure off Britain's ally, Frederick II, the Great, of Russia, but also to conform to a political agenda focused on demonstrating that the government was pursuing national interests and not simply sending troops to Germany. **ABOVE**

ANGLO-FRENCH NAVAL CLASHES, 1759 This illustration captures the tempo of operations and the far-flung nature of the struggle. British success was not due to superior weaponry as the ships and equipment were substantially the same. Instead, the crucial factors were a level of continuous high commitment and expenditure, an effective use of warships within the constraints of naval warfare and technology, and the inculcation of an ethos and policy that combined the strategic and operational offensive with tactical aggression. The British preferred to conduct the punishing artillery duels of the line-of-battle engagements at close range in contrast to the French preference for long-range fire. The key victories of 1759 were those of Lagos (Portugal) and Quiberon Bay which wrecked the French invasion plan and devastated their fleet. At Quiberon Bay on 20 November, Edward Hawke made a bold attack on the French fleet which had taken refuge, counting on the bay's shoaly waters and strong swell to deter the British fleet. In a ferocious wind, Hawke sailed boldly into the bay and in a confused engagement, British gunners and seamanship proved superior, with seven French ships of the line captured, sunk or wrecked. RIGHT

ANNO 1759.

...TERNEHMUNGEN zur SEE auf denen Französischen Küsten desglei... ...a in AFRICA und dem Mitternæchtlichen AMERICA aus denen beste... ...gen, u. nach denen besten Geograph. und Topographischē Karten, mit beobach... ...tē Gegenden, gezeichnet, u. mit Historischen Anmerckungē erläutert durch C.F.v.H.

CARTE des ENTREP: ANGLOISES par MER sur les Côtes de France aussi dans les Colonies Françoises en AFRIQUE et en AMERIQUE Septentrionale tirées des meilleures Relations et Avis speciels, et suivant les meilleures Cartes Geographiques et Topographiques, avec les Observations de Longitude et Latitude, de Signée et expliquée avec des Remarques Historiques par C.F. de H.

SEE-TREFFEN zwischen der Esquadre des Englischen Admiral Osborne von 14 Kriegs Schiffen u. der Französ: Esquadre les 3. Kriegs Schiffen und 3 Fregatten am 28. Febr: 1758. im Gesicht von Carthagena vorgefallen.

INS: CAPITIS VIRIDIS. I. S. Antonii, I. S. Vincentii, I. S. Lucia Branca, I. S. Nicolii, I. Sal, I. Bona Vista, I. de Mayo, I. S. Iago.

Englischer Überfall und Nachlassung d. 6 Juny 1758 zu S. Mallō ein Hafen in Nieder Bretagne mit einer Flotte von 37 Kriegs Schiffen unter dem Admiral Haucke, und 20 Transport Schiffen mit 16000 Mann unter Comando des Hertzogs von Marlbourg.

SARA DESERT, REGNUM, REG: GIALOF, REG: GUIAN, REG: SALUM, REGN: GENEGRES.

PRISE de LOUISBOURG et Cap Breton par l'Admiral Anglois Monsr: de Bossowen le 26 Juillet 1758.

Eroberung der Stadt und Festung LOUISBOURG und Insul Cap: Breton nebst St. Jean unter dem Englischen Admiral Bossowen den 26 Juliij 1758.

Überfall und Landung derer Engelländer zu CHERBURG einen Französischen Stadt und Hafen in der Nieder Normandie gerade gegen der Insul Wicht und Portsmuth, über geschehen den 7. August: 1759.

LA MANCHE ou LE CANAL.

SURPRISE Angloise de Cherbourg en Normandie le 7 Aout 1758.

PRISE de FRONSENAC en Amerique par les Anglois le 27 Aout 1758.

ACTION près CAST.

ACTION bey St. CAST.

PAIS DES ESQUIMAUX.

Eroberung der Insul ANTICOSTI und des Landes GASPE durch den Englische Admiral Hardy im Sept: 1758.

PRISE de l'Isle ANTICOSTI et Gaspe par l'Admiral Hardy au Mois Sept: 1758.

DU FLEUVE St. LAURENT.

St. Malō.

Gravé par J. A. Frederic et se vend chez lui meme à Augsbourg.

FIGHTING A LAKE BATTLE, 1776 The British followed up their success in clearing Canada of American invaders in May and June 1776 by cautiously advancing south to Lake Champlain. This advance revealed a facet of their capability because, on 11 and 13 October 1776, on the lake near Ile Valcour, a British flotilla that had been built from scratch under the command of Captain Thomas Pringle defeated an American flotilla under Benedict Arnold, destroying eleven American ships. RIGHT

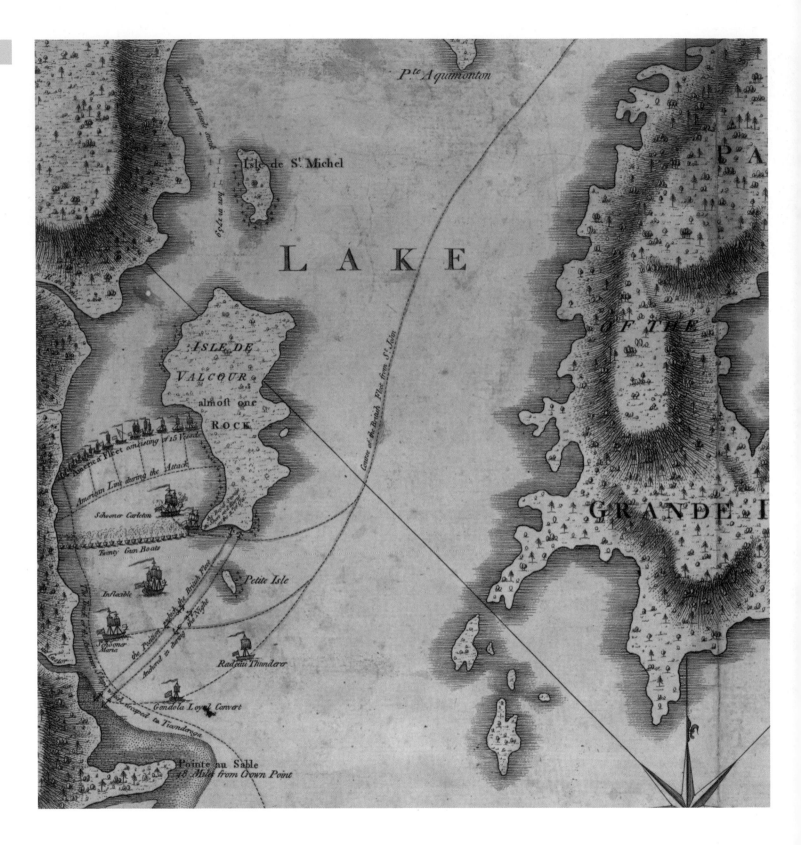

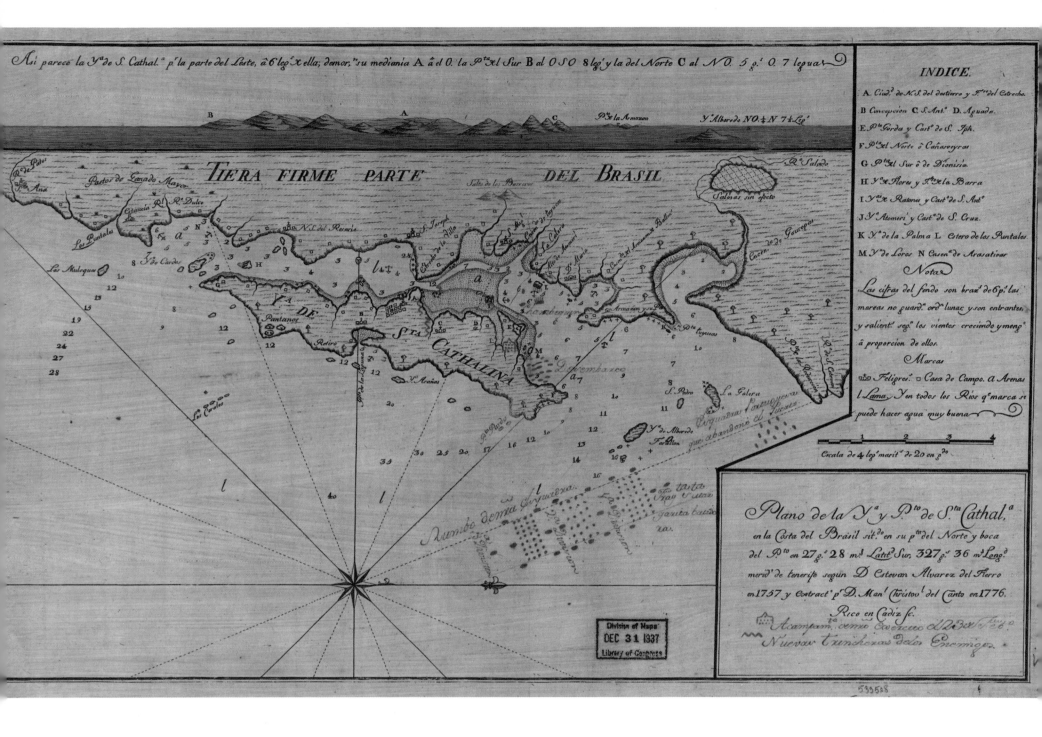

BATTLE OF SANTA CATARINA, 1777 Long-standing Portuguese-Spanish tensions over the area of modern Uruguay (notably the Portuguese base at Sacramento), led to a major Spanish naval expedition that attacked Santa Catarina island. The map shows this expedition, with the ships, which included troop transports, marked in red. Portuguese warships were brushed aside and the Portuguese positions were rapidly taken. The subsequent agreement left Sacramento as Spanish but not Santa Catarina. The Spaniards had also taken Port Egmont, the British base on the Falkland Islands, in 1770, but British naval pressure led the Spaniards to restore Port Egmont. **ABOVE**

MAPPING NAVAL WARFARE

NEWPORT, RHODE ISLAND, DURING THE

AMERICAN WAR OF INDEPENDENCE, 1775–83

The map shows fortifications, fields of fire, and positions of naval vessels. A French expeditionary force of 5,200 troops under the Count de Rochambeau, accompanied by seven ships of the line, anchored at Newport on 11 July 1780. General Sir Henry Clinton, the British commander, pressed Admiral Marriot Arbuthnot, a difficult man with whom his relations were increasingly poor, for a joint attack designed to destroy the French force, which he hoped would end the French threat. Obstruction on the part of the admiral prevented the mounting of an attack during the crucial period before the French could fortify Newport, Arbuthnot, for example, refusing to lend cannon. The following month, in turn, Clinton blocked Admiral George Rodney's plan for a joint attack on New York by offering too few troops. Disagreements between generals and admirals reflected the lack of an integrated command structure, but also arose from the very different interests of the two services. **RIGHT**

PLAN DE LA VILLE, PORT ET RADE DE NEWPORT, AVEC UNE PARTIE DE RHODE-ISLAND OCCUPÉE PAR L'ARMÉE FRANÇAISE AUX ORDRES DE Mr. LE COMTE DE ROCHAMBEAU ET DE L'ESCADRE FRANÇAISE COMMANDÉE PAR Mr. LE Chr. DESTOUCHES.

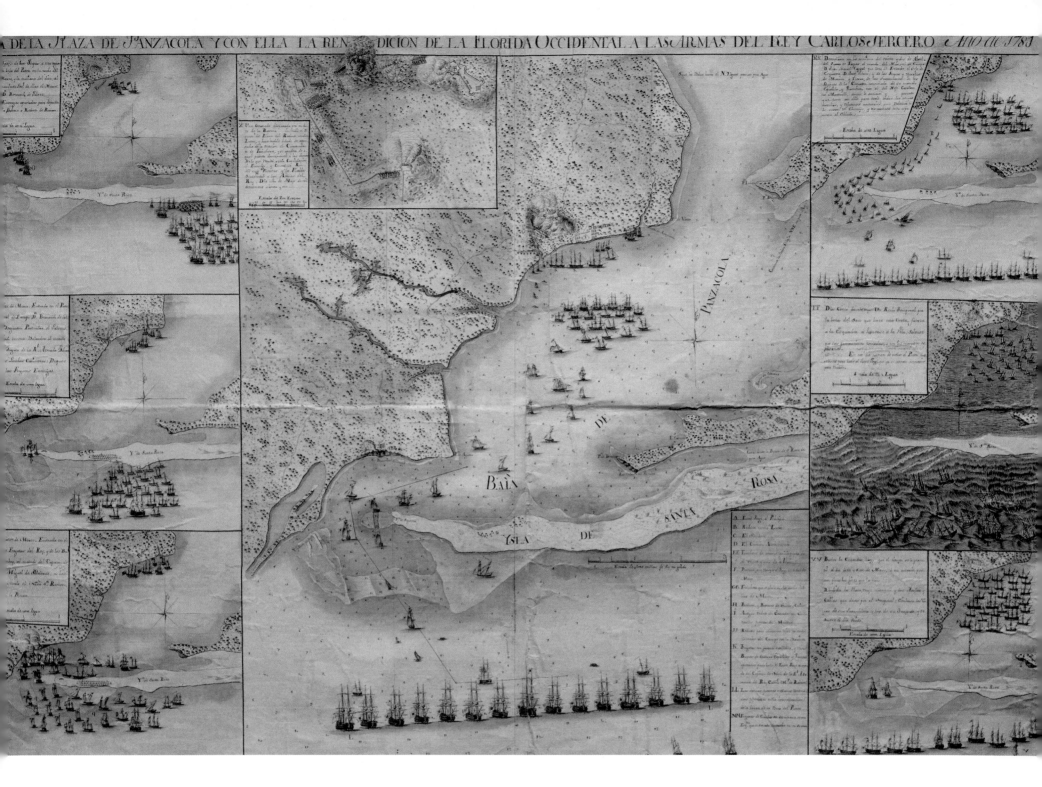

THE CONQUEST OF WEST FLORIDA, 1781 Spain's successful siege of Pensacola brought to an end a dazzling successful campaign in the Gulf of Mexico, one in which the Spanish Governor of Louisiana, Bernardo de Galvez, was backed by the Havana-based Spanish Caribbean fleet. Galvez captured the British forts at Manchac, Natchez and Baton Rouge in September 1779 and Mobile in March 1780. In May 1781, Pensacola fell to a far larger Spanish force. The defeated British commander attributed the defeat 'to the notorious omission or neglect, in affording Pensacola a sufficient naval protection and aid,' a warning about the danger to the British position elsewhere, as was seen at Yorktown later in the year. The Spanish navy had grown considerably in strength since 1763 and the Spaniards exploited their naval strength in the Gulf to move troops from Cuba and mount amphibious operations. Spain gained East and West Florida under the 1783 peace treaty. **ABOVE**

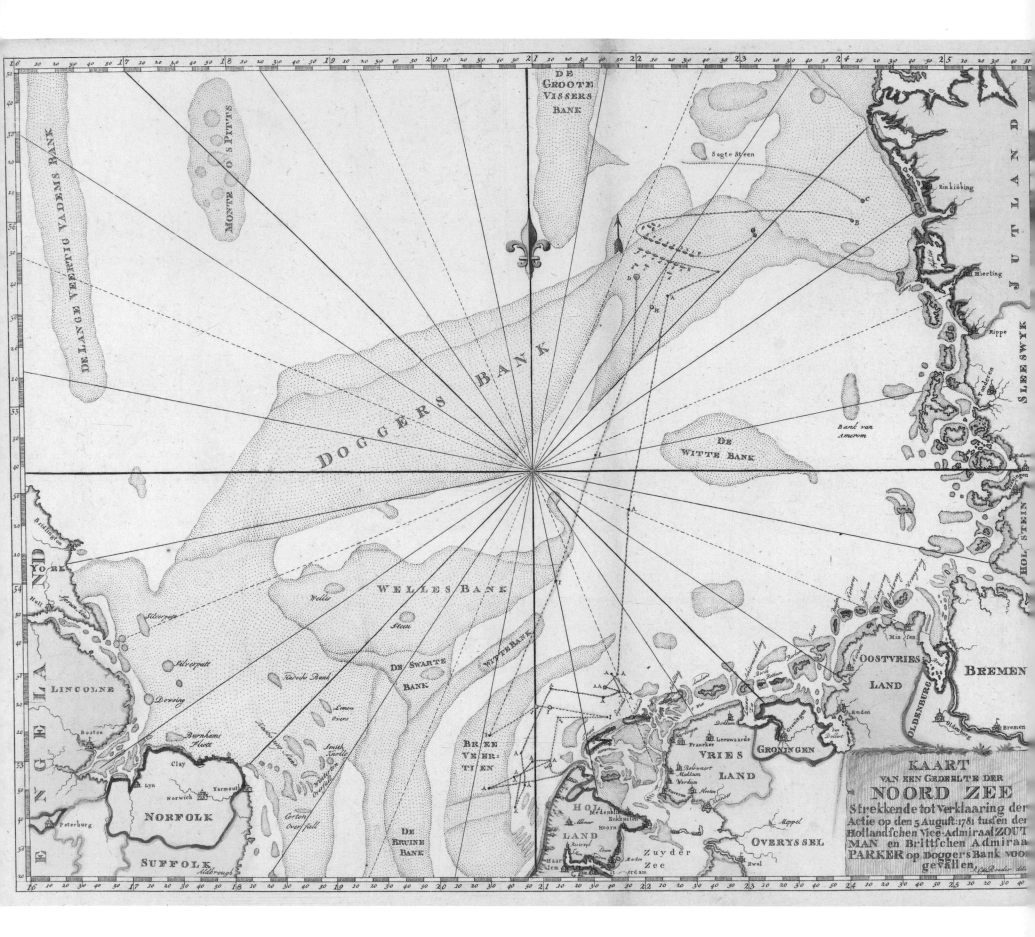

KAART
VAN EEN GEDEELTE DER
NOORD ZEE
Strekkende tot Verklaaring der
Actie op den 5 Auguſt 1781 tuſſen den
Hollandſchen Vice-Admiraal ZOUT
MAN en Brittſchen Admiraa
PARKER op Doggers Bank voor
gevallen.

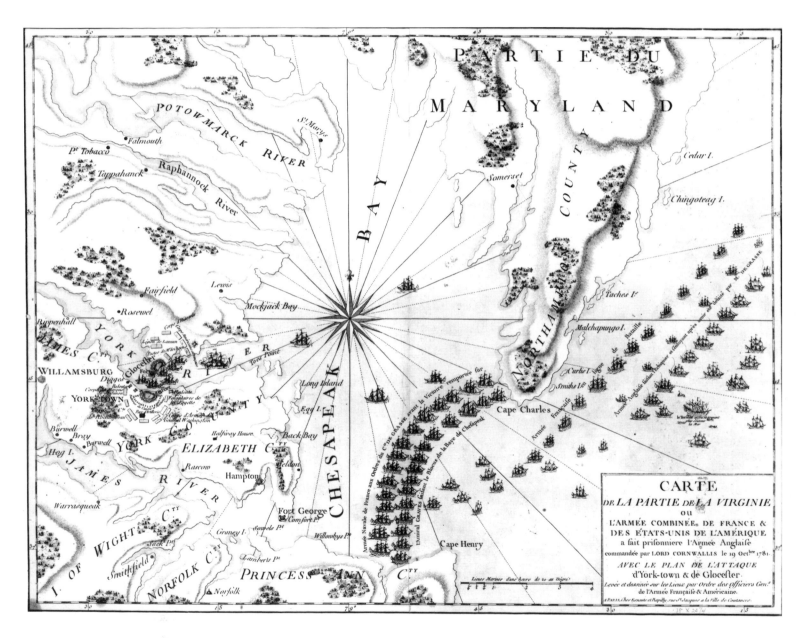

CHRISTOPH DE ROEDER The Dutch entered the war in 1780. On 5 August 1781 their fleet, under Rear-Admiral Johan Zoutman, fought in battle with the British squadron under the elderly Vice-Admiral Hyde Parker. Both were escorting a convoy of merchantmen. This map captured the location and moves of the warships in the North Sea. This was a costly, close-range, but indecisive, battle between two nearly equal fleets, each fighting in a fairly rigid fashion. The British suffered from having several old ships of poor seaworthiness. The battle had the strategic consequence of obliging the British to keep a squadron busy watching the Dutch. **OPPOSITE**

SIEGE OF YORKTOWN AND BATTLE OF THE VIRGINIA CAPES, 1781 The failure of Admiral Thomas Graves to defeat the French off the Virginia Capes was indecisive in terms of the damage inflicted, with neither side losing any ships. However, as it prevented the relief of the British army besieged at Yorktown, this was a major British defeat. Graves was outnumbered, nineteen to twenty-four in ships of the line. Instead of taking the risky course of ordering a general chase on the French van as it sailed in disordered haste from Chesapeake Bay, Graves manoeuvred so as to bring all his ships opposite to the French line of battle, which was given time to form and thus thwart him **LEFT**

PACIFIC CONFRONTATION, 1791 This nautical chart shows Port Hunter, now known as Balanawang, and Waterhouse Cove on Duke of York Island, Papua New Guinea. An ancillary view is of a Dutch transport ship carrying the crew of a British warship wrecked on a coral reef and firing on native canoes that attacked the transport ship in 1791. RIGHT

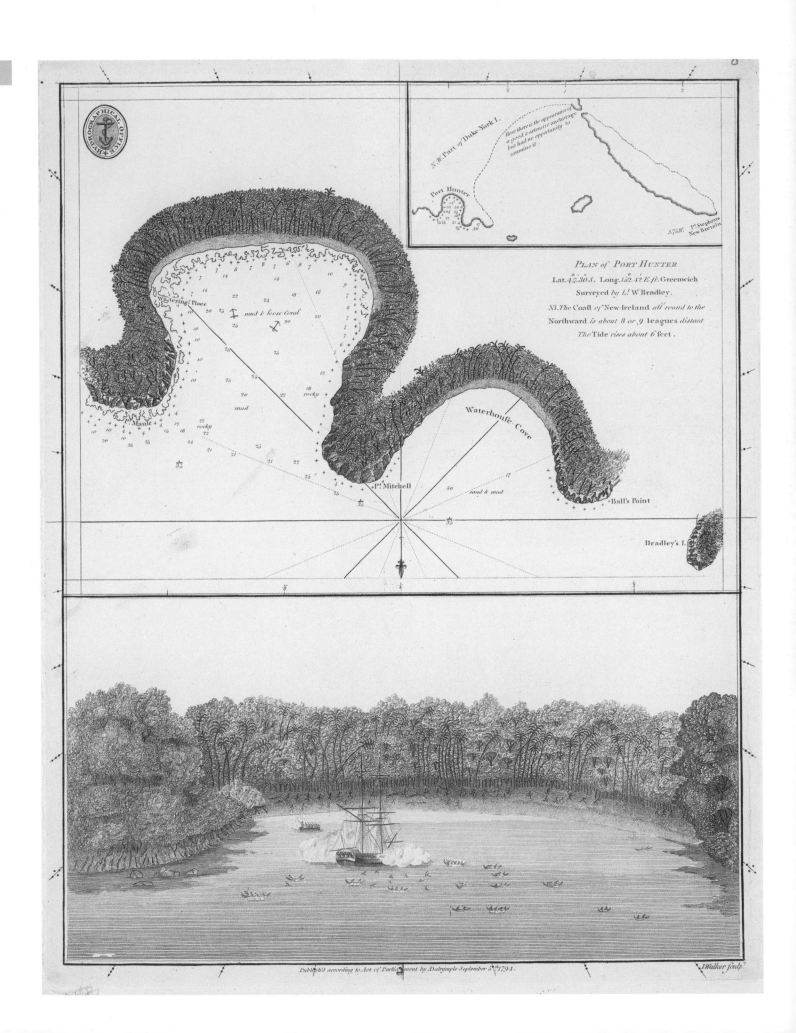

PLAN of PORT HUNTER
Lat. 4°.30 S. Long. 152.42 E.fr. Greenwich
Surveyed by Lᵗ W Bradley.
N3. The Coast of New-Ireland all round to the
Northward is about 8 or 9 leagues distant
The Tide rises about 6 feet.

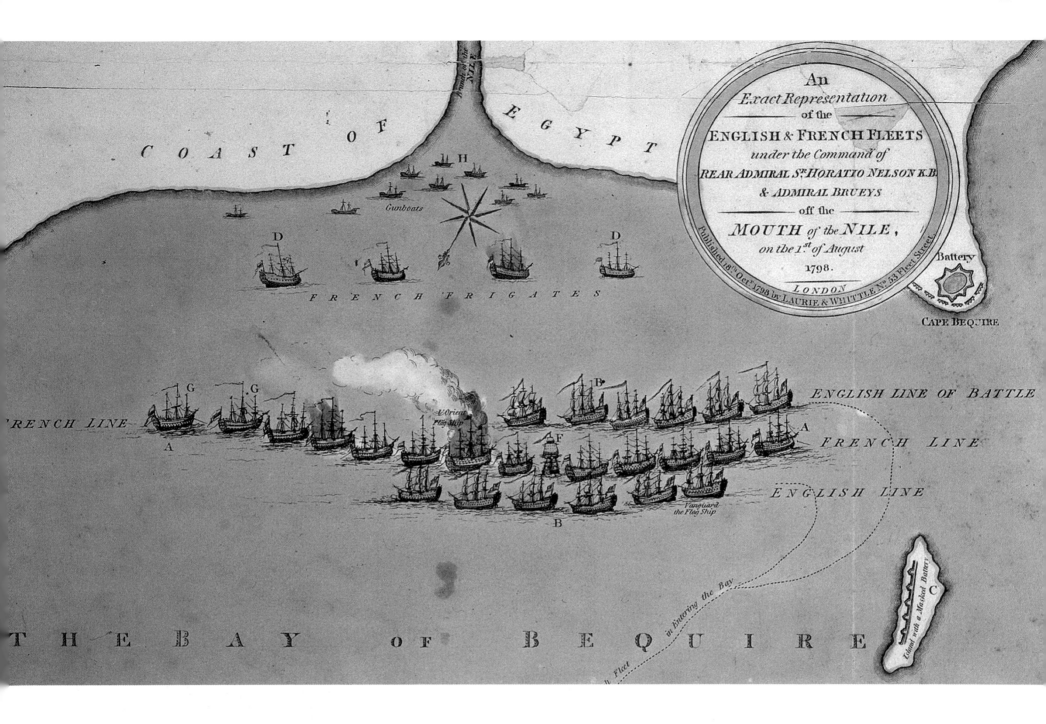

COAST OF EGYPT

An
Exact Representation
of the
ENGLISH & FRENCH FLEETS
under the Command of
REAR ADMIRAL Sr HORATIO NELSON K.B.
& ADMIRAL BRUEYS
off the
MOUTH of the NILE,
on the 1st of August
1798.
LONDON
Published 18th Oct.r 1798 by LAURIE & WHITTLE No 53 Fleet Street

Gunboats

D   D   D

FRENCH FRIGATES

Battery

CAPE BEQUIRE

H

G   G

B

L'Orient
Flag Ship

ENGLISH LINE OF BATTLE

FRENCH LINE

A

FRENCH LINE

A

F

ENGLISH LINE

Vanguard
the Flag Ship

B

THE BAY OF BEQUIRE

Island with a Masked Battery

C

in entering the Bay

n Fleet

BATTLE OF THE NILE, 1 AUGUST 1798 At dusk, Nelson unexpectedly attacked the French on both sides of their anchored line: on the shallow inshore side of their line, where the French were not prepared to resist, as well as simultaneously on the other side, a manoeuvre that was not without risks: HMS *Culloden* ran aground and was unable to take part in the battle. Fighting at night and with the British firing at very close range, the French lost eleven of their thirteen ships of the line present: the other two fled, as did the French frigates. The nature of the French position was such that Nelson was able to achieve a battle of annihilation, first defeating the ships in the French van and then pressing on to attack those moored behind; the latter had been unable to provide assistance. French gunnery proved inadequate, and the French were not only poorly-deployed but also failed to respond adequately to the British attack. The British navy worked as a well-integrated force. Nelson had ably prepared his captains to act vigorously and in co-operation in all possible eventualities and had fully explained his tactics to them. **ABOVE**

MAP OF THE BATTLE OF COPENHAGEN, 1801

The Danish ships at anchor in the harbour are indicated in red while the attacking British force is shown in blue. OPPOSITE

The nineteenth century saw Western powers become increasingly dominant on the world stage, with particular impact in Asia, Africa and Australasia. This process was related to their use of the information they could gather, deploy and apply, while, in turn, this usage reflected the potential offered by their power, and notably their maritime strength. This relationship was seen not only in political effectiveness, but also in other aspects, including economic activity. There were also attempts to advance a cultural dominance over non-Western peoples, and to frame, discuss and use information accordingly.

CHARTING THE OCEANS

The charting of the oceans was a key instance of this process and theme of the nineteenth century, one that brought together the search for information, its accumulation, depiction and use. This process was linked to power, as Britain's global commitments and opportunities, naval and commercial, made it both easiest and most necessary for Britain to acquire and use the information. Indeed, throughout the century, Britain played the major role in charting the oceans. In 1808, while Britain was at war with Napoleonic France, and very much dependent on its navy, the Charting Committee of naval officers was appointed by the Admiralty to advise on how best to improve the situation. This provided an instructive instance of the manner in which state agencies took on new functions and, in doing so, transformed the amount and reliability of available information. The committee recommended the provision of a set of charts for each British naval station as rapidly as possible and to cease relying on the private sector.

This entailed the purchase of charts from commercial map publishers and the buying back of the copyright of charts made by naval officers and produced by entrepreneurs. The latter practice, which had been very important in the eighteenth century, was symptomatic of the close relationship between the state and commerce in British map publishing. Underlining the complementarity sought between institutions acquiring information, there was also an effort to obtain information on the British coastline from the Ordnance Survey, which was the branch of the British military responsible for the mapping of the land. The coastline was a key zone as it was there that an invasion would land and where it would be resisted if earlier opposition had been unsuccessful at sea.

An atlas of *Charts of the English Channel*, containing thirty-one charts for naval use, appeared in 1811. Earlier charts were checked; for example, in 1797 the British checked the Dutch charts of the Molucca Islands in the East Indies. This proved necessary as the British conquered them during the Napoleonic Wars. Running aground was a major problem for warships mounting inshore operations, such as supporting amphibious attacks and enforcing blockades. The Royal Navy was particularly prone to losing ships this way as it frequently mounted such operations and was the largest navy in the world. The British lost more ships running aground than from enemy action. Many such losses occurred in European waters, in large part because of the British blockade of France and its allies there, but other waters also saw significant losses, including the Indian Ocean.

After the Napoleonic Wars ended in 1815, the British pressed on to chart coastal waters across the world. This was part of the process in the West by which formal, state-directed, information-gathering replaced earlier, *ad hoc* means of assembling information that had been gathered in a non-systematic fashion. Much of the existing situation was in practice unsatisfactory. Robert Sawyer reported of

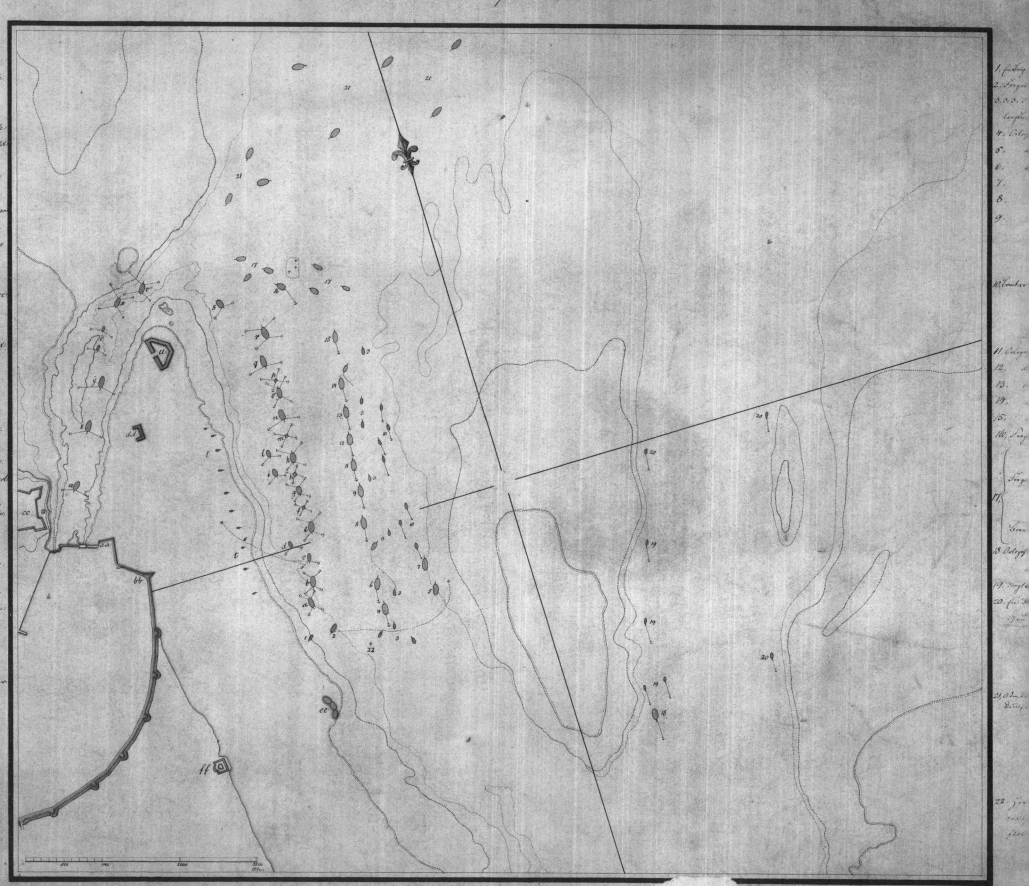

Plan over

Slaget paa Kjöbenhavns Rhed

den 2den April 1801.

THE BATTLE OF COPENHAGEN, 1801 Facing a deteriorating diplomatic and military situation, Britain, in 1801, took action against the threatening Northern Confederacy of Baltic powers. Denmark rejected an ultimatum to leave the confederacy. At the battle of the Nile in 1798, Nelson had been victorious at the expense of an anchored line of French warships. At the battle of Copenhagen, on 2 April 1801, Nelson was again successful at the expense of an anchored line. After sounding and buoying the channels by night, he had sailed his division down the dangerous Hollaender Deep in order to be able to attack from an unexpected direction. Heavy Danish fire led Nelson's commander, Sir Hyde Parker, to order him to 'discontinue the action', if he felt it appropriate, but Nelson continued the heavy bombardment and the Danish fleet was battered into accepting a truce. The seventeen Danish ships of the line that were present were captured or destroyed. Neutral trade with France was thus no longer an option. Nelson's reputation rose, *Bonner and Middleton's Bristol Journal* reporting on 25 April 'The zeal, spirit, and enterprise of Lord Nelson were never more completely developed than upon this great and memorable occasion, and they happily diffused their influence through the whole of the squadron under his immediate command.' These show different ways in which the battle could be represented and mapped. RIGHT

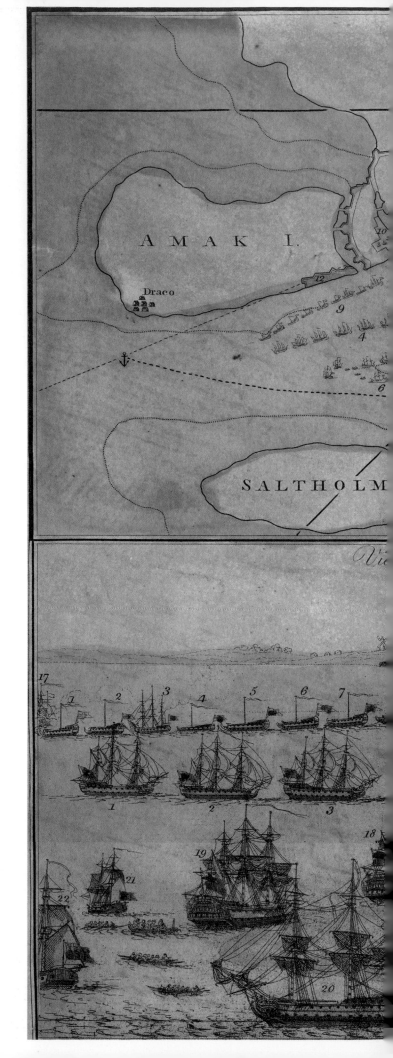

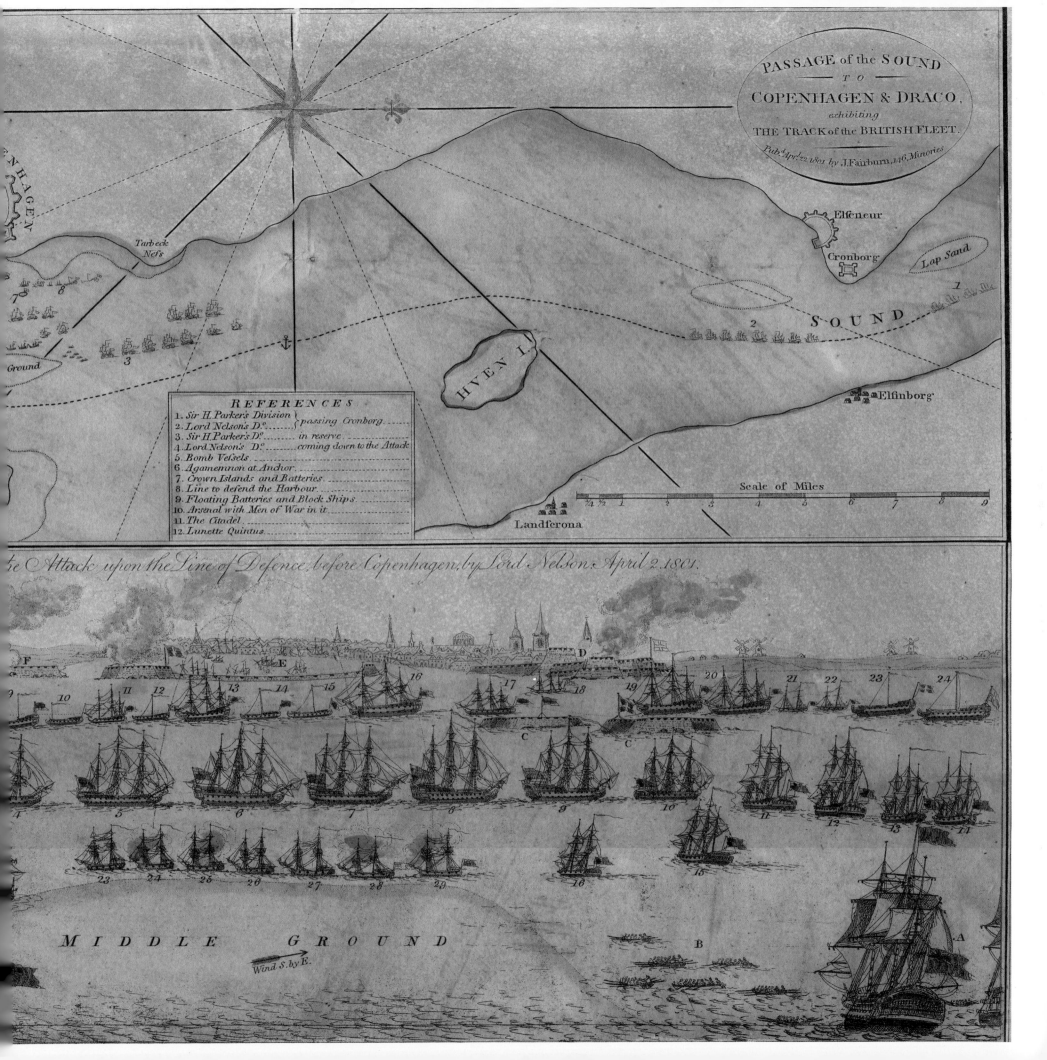

PASSAGE of the SOUND
TO
COPENHAGEN & DRACO,
exhibiting
THE TRACK of the BRITISH FLEET.

Pub.d Apr.l 22. 1801 by J. Fairburn, 146, Minories.

Elfeneur

Cronborg

Lap Sand

SOUND

HVEN I.

Elfinborg

ENHAGEN

Tarbeck Nefs

Ground

REFERENCES
1. Sir H. Parker's Division }
2. Lord Nelson's D.o } *passing Cronborg.*
3. Sir H. Parker's D.o .... *in reserve*
4. Lord Nelson's D.o .... *coming down to the Attack*
5. Bomb Vefsels
6. Agamemnon at Anchor
7. Crown Islands and Batteries.
8. Line to defend the Harbour.
9. Floating Batteries and Block Ships
10. Arsenal with Men of War in it
11. The Citadel
12. Lunette Quintus

Scale of Miles

Landferona

Attack upon the Line of Defence, before Copenhagen, by Lord Nelson, April 2. 1801.

F   E   D   20   21   22   23   24
9   10   11 12   13   14   15   16   17   18   19

C   C

4   5   6   7   8   9   10   11   12   13   14

23   24   25   26   27   28   29   15
                                                16   B   A

MIDDLE   GROUND

Wind S. by E.

the Sunda Strait, a major route through the East Indies, 'till the extent of the dangers off the south end of Banca are better known, the approach of it must be dangerous and we seem to be equally ignorant of what dangers may lie off the numerous islands to the south east'. In 1817, en route to Guangzhou (Canton), HMS *Alceste* hit a coral reef just north of Sunda.

Alongside key surveys, such as that of the Thames Estuary, much of the charting was far-flung. Moreover, much of it was not a question of charting coasts controlled by Britain. For example, a British coastal survey of 1822–24 brought back much information about East Africa. There was also a continuing relationship with the private sector in Britain, although the terms of exchange were now different to those of the eighteenth century. From 1821, the extensive range of charts produced by the Admiralty was offered publicly for sale, a policy designed to produce funds for more surveying. Indeed, catalogues of material for sale were published from 1825.

INDIVIDUAL CONTRIBUTIONS

The careers of the individual naval officers and ships responsible for such surveys reflected the range of British activity, but also the extent to which surveying was linked to British power. This situation looked towards the later dominance of Britain in submarine telegraphy, a dominance that was a matter not only of charting the oceans, ownership of telegraph lines and the related production and laying of cables, but also of the key role of Britain in setting standards.

Sir Francis Beaufort (1774–1857), whose father, the Reverend Daniel Beaufort (1739–1821), had published a map of Ireland in 1792, which proved useful at the time of the suppression of the 1798 nationalist rising, entered the Royal Navy in 1787 and received nineteen

wounds in 1800 when he cut out a Spanish warship. In 1807, on the pattern of James Cook's surveying of the St Lawrence prior to the British attack on Quebec in 1759, Beaufort surveyed the entrance to the Plate estuary. This was a valuable aid to the warships in preparing for a large-scale but unsuccessful British attack on Buenos Aires. More lastingly, knowledge of these waters would be important for what was to be a major destination for British trade.

As a frigate captain, Beaufort was active in 1810–12 in Turkish waters, seeking to suppress pirates and to survey the coast, only to be badly wounded in a clash. He subsequently produced charts based on his survey. Alongside William Smyth's hydrographic surveys, Beaufort's *Karamania* (1817) showed the process by which the British controlled the Mediterranean through naming it. Smyth published *The hydrography of Sicily, Malta, and the adjacent islands* (1823) and surveyed the Adriatic and the North African coast. He rose to be a rear-admiral, and became president of the Royal Geographical Society and the Royal Astronomical Society.

In 1829, Beaufort became hydrographer to the Navy, a post he held until 1855. Soon after his appointment, Beaufort plotted on a map of the world the coasts already covered by surveys. This was an aspect of the systematic approach to information gathering that became more important in the nineteenth century. Concerned by the length of the coastline not yet tackled, a length clarified by the map, Beaufort pressed on to fill the gaps.

The results were shown in the flow of information received. For example, Edward Belcher surveyed the coast of West Africa in the early 1830s, an area of major concern as Britain sought to suppress the slave trade. This was to be an important naval commitment during much of the century. William Fitzwilliam

Owen and Alexander Vidal surveyed the coastline of Africa from 1821 to 1845, Later in the 1830s, Belcher surveyed the west coast of South America, his surveying an aspect of Britain's important informal empire in Latin America. Trade was helped, and British naval power acquired not only practical knowledge but also considerable prestige.

Robert Moresby surveyed the Red Sea in 1829–33. British warships operating there in 1799 against the French in Egypt had run aground. This was a classic example of charts playing a role in the negotiation of treacherous waters for trade and power projection. As an instance of the cumulative nature of change, the impact of the new charts of the Red Sea, published in

1834, was enhanced by the development of steam navigation, which enabled ships to overcome both northerly winds in the Red Sea and calms in the Mediterranean. This capability demonstrated the flexibility of steam and its multi-purpose value.

In turn, power considerations came into play, with Britain determined to control the shortest route to India. This represented a new strategic responsibility and new operational tasks. The army was ultimately dependent on the navy's ability to support a greater range of commitments and a large number of positions. Mauritius was conquered in 1810, Aden was annexed in 1839 and control was taken over the Suez Canal soon after it opened in 1869. Cyprus followed in 1878 and Egypt, with its major naval base at Alexandria, in 1882.

The quest for accuracy was also seen in the Royal Navy's patronage of Charles Babbage, a key figure in the development of computing. The navy wanted astronomical tables without printing errors, which would allow it to plot positions with certainty. Originally Babbage's difference engine was to have a tool for moulding *papier mâché* type that could be used for printing, thereby cutting out human error in typesetting.

USE OF CHARTS

Charts were extensively used in war. The British capture of Mauritius from the French in December 1810 was preceded by careful charting of the waters round the island. Earlier in 1810, in a major blow that reflected a lack of information, two frigates from the British blockading squadron had run aground and been destroyed. Similarly, the collection of the Royal United Services Institute (of London) includes a *Chart of the island of Chusan enlarged from a chart by Alexander Dalrymple* [hydrographer to the Navy, 1795–1808], *and corrected in many places by observations made during an expedition under Captain Sir H Le Fleming Senhouse on HMS Blenheim between 30th August and 4th September 1840.* This information was both used and improved in Britain's First Opium War with China, a conflict that also led to the charting of the seas round Hong Kong in 1841 by HMS *Sulphur*, under the command of Edward Belcher. Moreover, Britain's new presence in China led to an extension of information on the region. As commander of HMS *Samarang* in 1842–47, Belcher surveyed the coasts of Borneo, the Philippines and Taiwan, the last then a part of China. This surveying reflected the new-found projection of British power, notably into northern Borneo.

Charts were not only used in war. They also helped in understanding the opportunities offered by the oceans. Information about the availability and distribution of whales, seals and fish led to an expansion of maritime activity. In time, however, this expansion hit the sustenance of local people, for example, the Yamana of Tierra del Fuego. More famously, British charting voyages contributed to the development of the theory of evolution. Charles Darwin found his voyage of 1831–36, as naturalist on HMS *Beagle*, a formative experience, particularly the journey to the Galapagos archipelago. This was a formative period akin to that for Joseph Banks on James Cook's HMS *Endeavour*, in the Pacific.

The captain of HMS *Beagle*, Robert FitzRoy, was a prominent figure in the development of meteorological services, becoming, in 1854, chief of the Meteorological Department of the Board of Trade, a post he held until 1865. FitzRoy placed a great reliance on barometers and helped to design an inexpensive one. Writing on how best to use barometers, he founded in 1861 a system of storm warnings, which became the basis of what he called the 'weather forecast'. The first telegraphic weather reporting was carried out in 1865. As an instance of the continuing process of nomenclature, the sea area Finisterre was renamed FitzRoy in 2002.

SURVEYS BY OTHER NATIONS

States other than Britain also charted waters, both their own and those elsewhere in the world. Under Charles-François Beautemps-Beaupré, between 1815 and 1838, the Atlantic coastline of France was mapped, enabling an application of triangulation by France that war had cut short over the previous half-century. Spain might have been a declining imperial power and lacked a powerful maritime base, but the *Dirección*

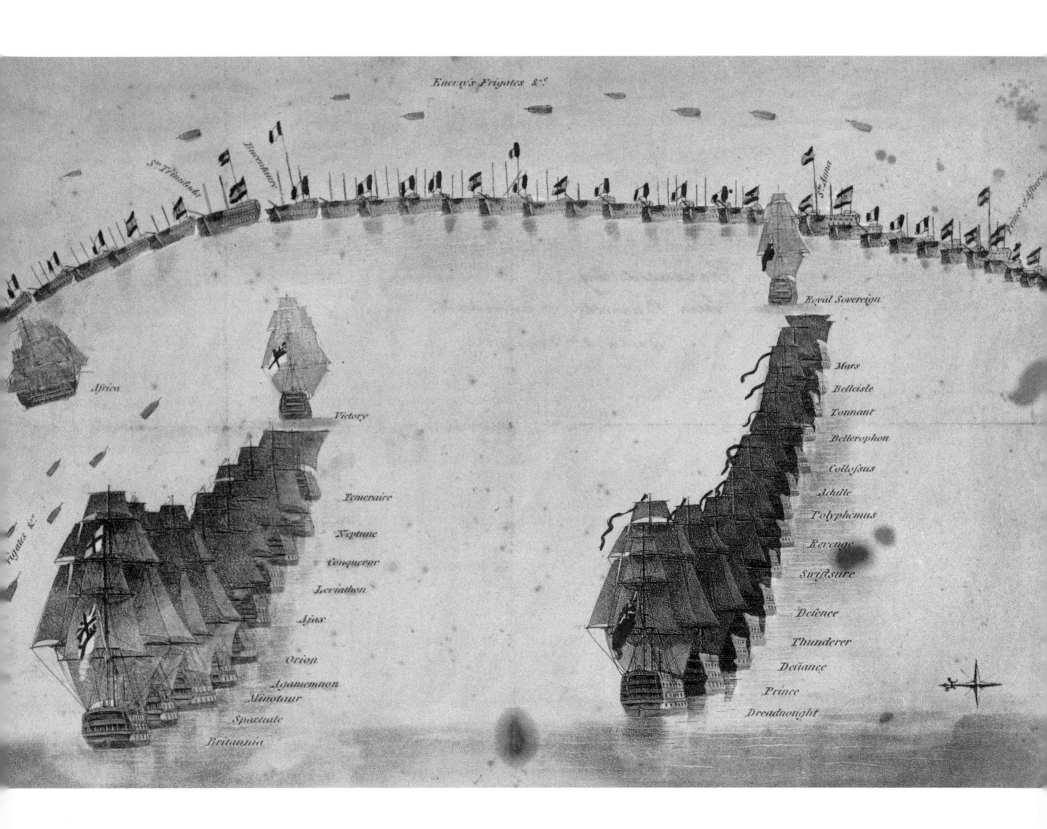

Enemy's Frigates &c.

Ste Trinidada   Bucentaur   Ste Anna   Prince & Asturias

Royal Sovereign

Africa

Victory   Mars

Belleisle

Tonnant

Bellerophon

Colossus

Temeraire   Achille

Neptune   Polyphemus

Conqueror   Revenge

Leviathan   Swiftsure

Ajax   Defence

Thunderer

Orion   Defiance

Agamemnon   Prince

Minotaur   Dreadnought

Spartiate

Britannia

Frigates &c.

BATTLE AT RATAN, 1809 BY, CARL GUSTAF GILLBERG

The last battle on Swedish soil, which
took place on 19–20 August 1809. Having
conquered Finland from Sweden, the
Russians had a small force in the northern
parts of what is now Sweden. The Swedes
sought to eliminate this through a
combined assault from land and sea,
but the Russians moved more rapidly,
defeating the Swedish force landed at
Ratan at Sävar on 19 August. The next day,
the Swedish force was attacked when
evacuating from Ratan but the artillery
fire from Swedish warships shown in the
illustration kept the Russians at bay.
Peace followed soon after. OPPOSITE

de Trabajos Hidrográficos (Spanish Hydrographic Office),
established in 1800, had active regional branches in the
Spanish empire. They mapped Spain's still-extensive
colonies, notably the north coast of Cuba in the 1860s
and Puerto Rico in the 1890s, although were less active
in the very extensive Philippine archipelago.

The USA built up considerable expertise in the
sphere of surveying, and it was a significant adjunct of
American power projection. It is important not to allow
a focus on the Royal Navy to overshadow such activity.
The United States Exploring Expedition of 1838–42 led
by the hyper-aggressive Charles Wilkes, was responsible
for surveying and oceanographic work, as well as power
projection in the Pacific. Wilkes publicised his work
with a multi-volume narrative. Copies of Wilkes' maps
were often the only items available for many parts
of the South Pacific during the Second World War.
In 1852–55, another American naval expedition, the
North Pacific Surveying Expedition, greatly expanded
hydrographic knowledge of Japanese waters.

There was also surveying in South America. In 1855,
USS *Water Witch*, a lightly-armed American naval
steamer which had ascended the Paraná river on a
mapping expedition, was fired on by Paraguayan forces.
This led, in 1858, to the dispatch of an American
squadron, which produced an apology, an indemnity
and a permission for the mapping expedition to
proceed. The United States Coast Survey proved a
major tool for the Union during the American Civil
War (1861–65), supporting its amphibious operations
and blockade.

Hydrographic offices were established in Russia in
1827, the USA in 1830, and in the Netherlands in 1856.

NAVAL POWER

Such information was necessary because of the
increasingly global range of the major naval powers.

This was a process that involved the use of strength,
but not necessarily conflict, at least with each other,
as opposed to with weaker powers. Indeed, after the
end of the Napoleonic Wars in 1815, the leading naval
powers did not fight again until 1914 when the First
World War broke out. In the meantime, there were
significant struggles, notably the Crimean War of
1854–56 and the Franco-Prussian War of 1870–71, in
which naval conflict only took a minor or secondary
role, although the French navy secured the flow
of material to sustain the resistance after the fall of
Napoleon III. In the Danish War of 1864, the Danish
navy prevented the Austrians and Prussians from
matching the pressure they were successfully exerting
on land. During this period, naval conflict could be
significant, as in the American Civil War (1861–65),
the Austro-Italian war of 1866 and even more in the
Spanish-American War of 1898.

Alongside these last three wars, in each of which
there were significant clashes, with warships sunk,
naval considerations were important in other conflicts,
even if there were no such clashes. This importance
varied greatly. For example, the presence of a major
naval power might deter intervention by other powers.
This was very much the case in the most significant
struggle in the New World, namely the Latin
American Wars of Independence in the 1810s and
1820s against Spanish and Portuguese rule. The
possibility of French-supported Spanish intervention,
in order to defeat or reverse independence, was very
much thwarted by the support for independence
provided by a benign British attitude. As Britain was
the leading naval power, this attitude precluded any
hostile intervention. British support, moreover, helped
in the development of local navies, notably that of
Chile. These navies played a major role in conflict
within Latin America, because, having gained

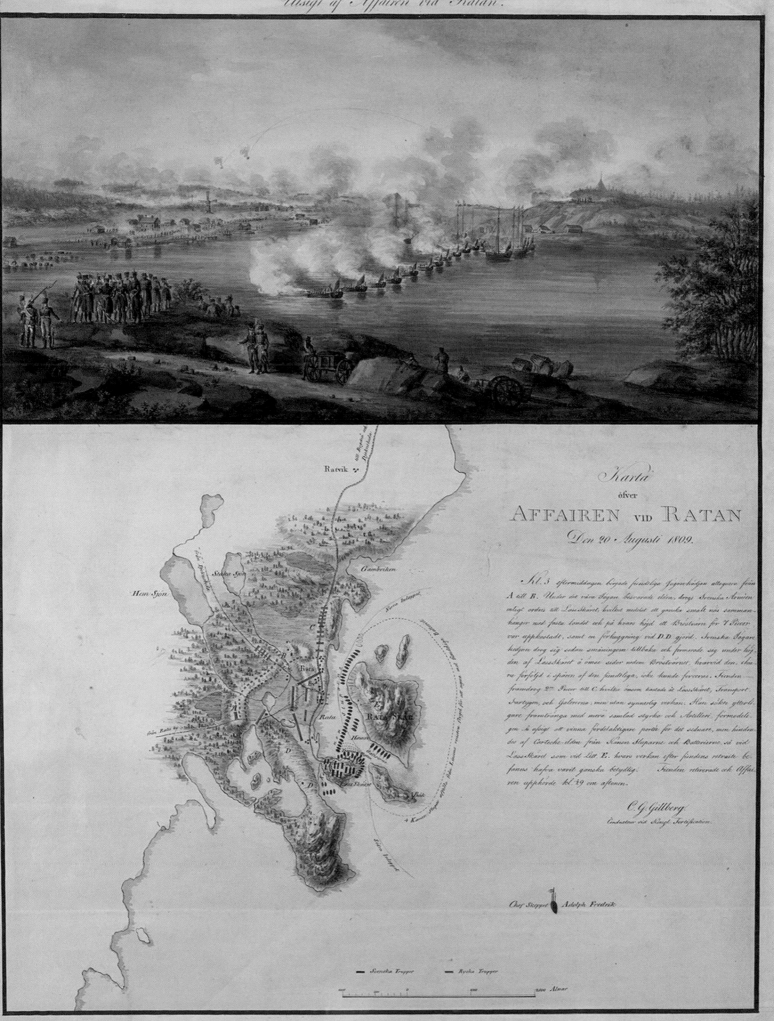

Utsigt af Affairen vid Ratan.

Karta öfver AFFAIREN VID RATAN Den 20 Augusti 1809.

BATTLE OF PLATTSBURG, 1814 The British invasion of the United States along the Lake Champlain corridor by about 10,000 troops hinged on local naval superiority. A British advance along the western shore of the lake had been delayed at Plattsburg while the British waited for a naval clash to determine whether they could use the lake to move their supplies. The American squadron under Thomas Macdonough was well-positioned and well-prepared, and on 11 September it fought well. Excellent American seamanship, good command decisions, and the strength of the short-range American cannonades enabled the Americans to compensate for British strength in long-range gunnery. All the major British ships were sunk or captured and the commander, Captain George Downie, was killed by an American cannonball. On the day of the battle, British land forces attacked the Americans south of Plattsburg, but the attack was abandoned after the defeat on the lake. **RIGHT**

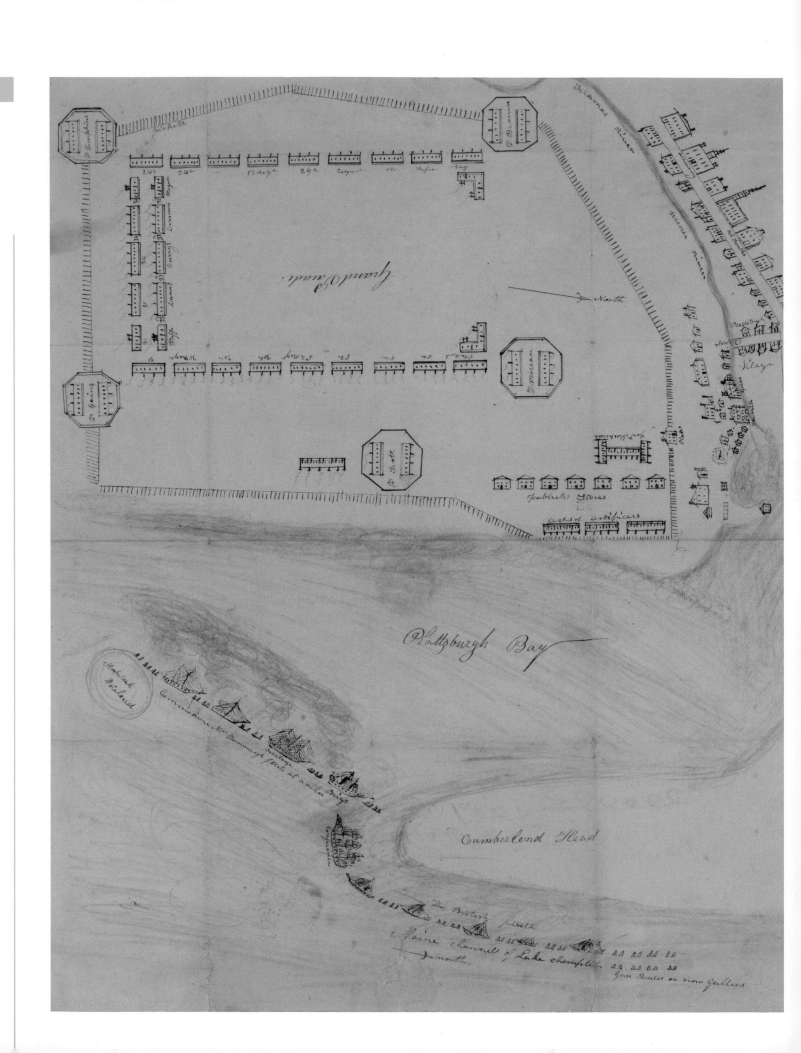

independence, the states both fought each other and faced rebellions. The display of naval power and the transport of troops were important, for example, in Brazil, where the road system, as more generally in Latin America, was terrible.

Similarly, the decision not to intervene in conflict could be crucial, for example, in the American Civil War (1861–65). Britain and France came close to intervening against the Union in 1861 and, even more, 1862. Had they done so, it is unclear what would have happened as far as combat was concerned: Union monitors (steam-powered iron warships that sat low in the water) and coastal artillery would have been able to inflict considerable damage, and Union warships that were commerce raiders would also have posed a threat. At the same time, the attacking power generally has the advantage of being able to concentrate strength at the point of initiative, and this would have been a formidable threat to the defences of Union coastal cities, particularly New York. Even if these had not been attacked, an offshore presence acting as a blockade would have been greatly disruptive.

This, indeed, was the experience of Russia during the Crimean War (1854–56) when the British presence in the Baltic Sea had this effect, with the possibility of attacks on St Petersburg, Riga and Tallinn. There were no naval battles during the Crimean War because the Russians chose not to send their navy out from its anchorages. That, however, did not mean that naval power was inconsequential. Indeed, Anglo-French power projection, notably to Crimea, would have been impossible without assured naval dominance. Moreover, this dominance made coastal bombardments and/or assaults possible, as at Kinburn on the Black Sea. The Crimean crisis itself indicated the varied uses of naval power, as the fate of the Black Sea was dramatised by the sweeping Russian naval victory over the Turks off

Sinope in 1853. Concern about the consequences of the battle encouraged Britain and France both to war with Russia and to making the Russian Crimean naval base at Sevastopol a key target.

Inadequate mapping affected Anglo-French naval action. As a result of the use of outdated maps, warships steamed into the wrong sea channels or ran aground. This affected the Allied operations in the Baltic Sea where the Bay of Bothnia was essentially uncharted.

WESTERN STRENGTH

The Russian victory off Sinope captured a central aspect of naval strength and warfare during the century, namely the superiority of Western states at sea over non-Western rivals. This process continued until Japanese naval victories over Russia in 1904–05, and these victories were obtained only thanks to Japan's Westernisation of its navy as a result of British influence and example. Nineteenth-century Western dominance was not generally captured in battle as non-Western naval forces were weak and frequently did not risk battle. Nevertheless, the British, notably the steamer *Nemesis*, defeated Chinese warships during the First Opium War, while the French proved similarly successful in the Sino-French War of 1884–85. The inability of the Filipinos to contest at sea the American takeover of the Philippines from 1899 was as significant. Naval warfare frequently involved power projection and punitive expeditions rather than full-scale battle.

To Western public opinion, notably that of Britain, a key element in the use of naval power was a moral component, particularly as Britain sought to end the slave trade and to stamp out piracy. The resulting operations and patrolling involved much naval activity and were important to the spread of empire, both

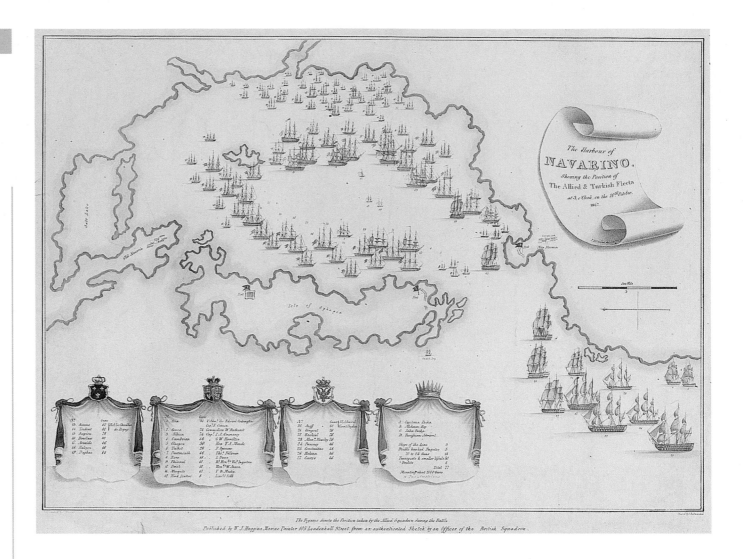

BATTLE OF NAVARINO, 1827 Thanks to the overwhelming British firepower of thirty-two-pounders at almost point-blank range, an Anglo-French-Russian fleet under Sir Edward Codrington destroyed the Ottoman and Egyptian fleets at the Battle of Navarino, a key event in the struggle for Greek independence. In the Treaty of London, Britain, France and Russia had decreed an armistice, only for this to be rejected by the Ottomans, and for the fleet to be instructed to enforce it. This battle, on 20 October 1827, was the last great one of the 'Age of Fighting Sail'. Western fatalities, 177, were far lower than the almost 17,000 of their opponents. The British government worried about Russian expansionism, but the public reacted with pleasure, and Codrington received chivalric orders in France, Greece and Russia. The map shows the position of the ships with a coloured key to indicate nationality. No soundings are included.

formal and informal. This activity involved much inshore work and, therefore, a need for accurate charts.

Some of the operations against what was defined as piracy were large-scale, as in 1819 when a British force of twelve warships, twenty transport vessels and 3,000 troops attacked Ras al-Khaimah in the Persian Gulf, an operation assisted by Omani ships. The cannon of the British warships played an important role in the breaching of the fortifications. Subsequently, British warships attacked strongholds along the Gulf in 1819–20, leading to treaties in which the local rulers agreed to accept the end of piracy and slave trading, although the last proved unenforceable. Britain decided to establish a naval base at Qais. In 1853, there was a permanent ban on maritime warfare in the Gulf in return for British protection. Thereafter,

British warships in the Gulf had a brief to 'watch and cruise'.

More generally, most such operations were small-scale and conducted by small ships, such as the British schooners on the coast of British Columbia which overawed Native Americans, but they were a key aspect of naval force in this period and an important basis for empire. Operations against pirates on the Malayan coast enabled the British to extend their influence way beyond their Straits Settlements – Penang, Malacca and Singapore – and, combined with these positions, gave Britain the dominant position on the route between the Indian Ocean and the Far East. The capability of the Royal Navy to conduct these small-scale operations indicated the capability and initiative of relatively junior commanders.

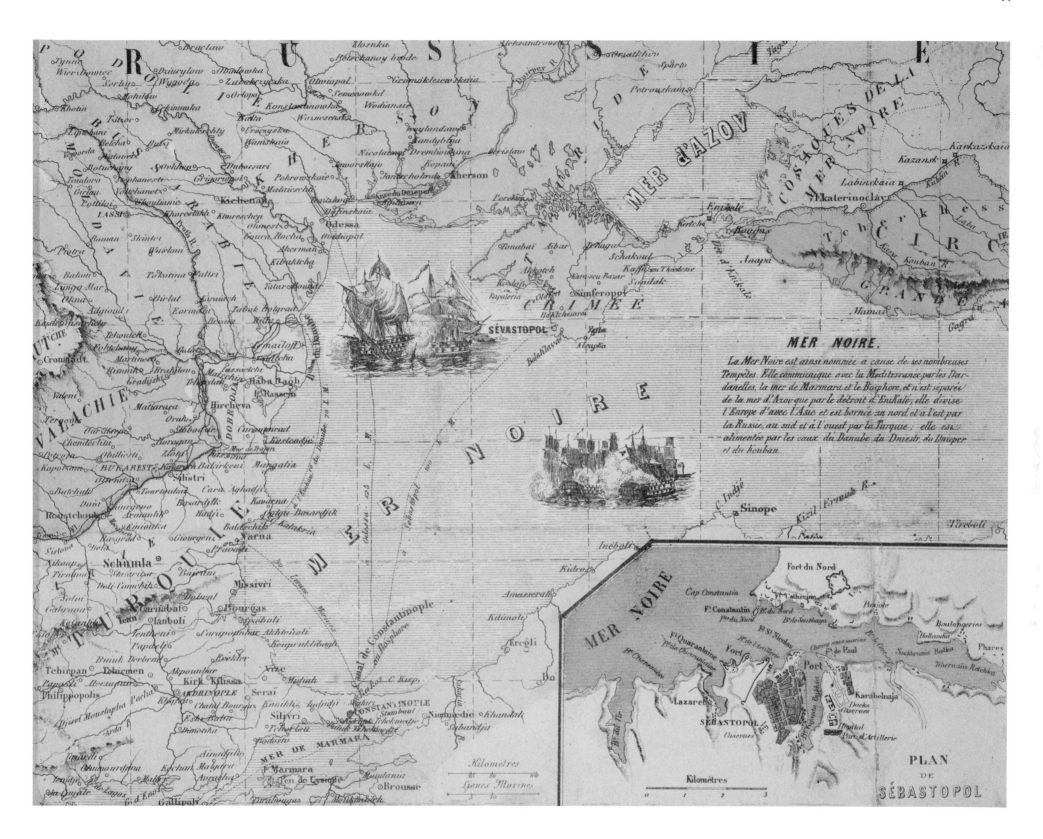

THE BLACK SEA, 1856 The destruction of a wooden Turkish squadron at Sinope on 30 November 1853 by a Russian fleet carrying shell-firing guns led to British and French intervention in 1854. This came to focus on Russia's Black Sea fortress port of Sevastopol. Because the Russians remained in port, there were no naval battles, while the Anglo-French naval bombardments of Odessa and Sevastopol in 1854 underlined the vulnerability of wooden warships to effective defensive fire by red-hot shot. As a result, with Sevastopol blockaded by mines, the focus moved to an invasion of Crimea, which led to a siege of the port. This map showed the information available to the French public. **ABOVE**

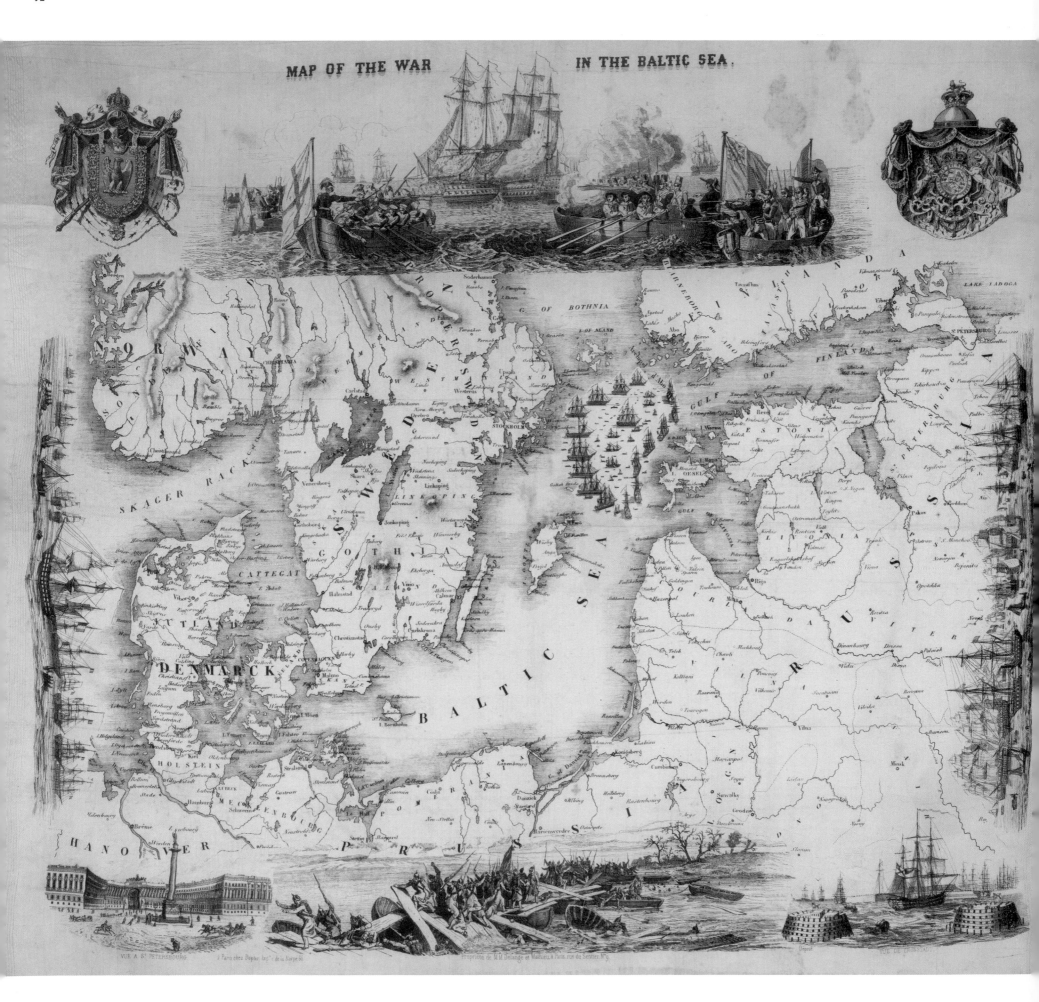

MAP OF THE WAR IN THE BALTIC SEA.

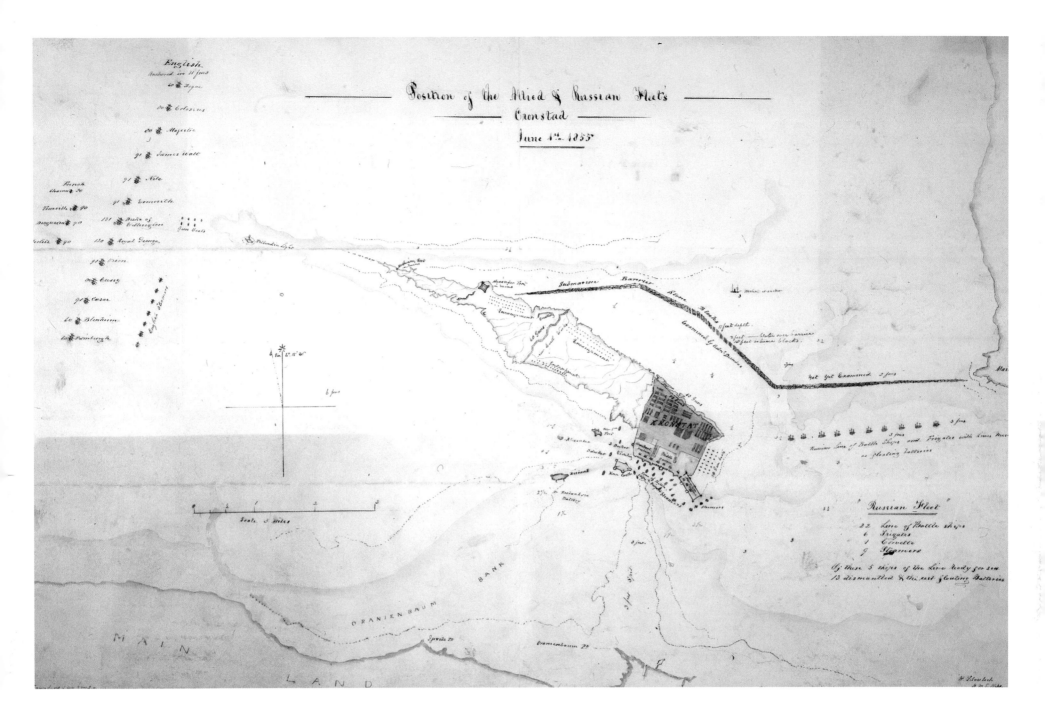

Position of the Allied & Russian Fleets — Cronstad — June 1st 1855

MAP OF THE WAR IN THE BALTIC SEA (THE CRIMEAN WAR), C.1855, PARIS, DOPTER A handkerchief map of the Baltic Sea during the Crimean War, decorated with vignettes of St Petersburg, Kronstadt, naval scenes and French and British coats of arms. The threat to St Petersburg may well have been an important factor in precipitating the end of the war. The ability of steamships to work with confidence close inshore affected the tactical, operational and strategic potential of naval power. British firepower increased. The first rifled gun to be used by navies was the Lancaster oval bore eight-inch gun. **LEFT**

POSITIONS AT KRONSTADT, 1855 A major British fleet was sent to the Baltic Sea during the Crimean War, but the outclassed Russians, based at Kronstadt off St Petersburg, refused to engage in battle. As a result, the British were able to engage coastal targets, notably Sveaborg, the fort that guarded the approach to Helsinki, although not to inflict decisive damage. The Russians mobilised a large number of steam-powered gunboats with heavy pivot guns to defend Kronstadt. It was to be attacked by British naval aircraft during the Russian Civil War. The map shows the positions at Kronstadt on 1 June 1855. **ABOVE**

THE SIEGE OF KINBURN, 1855 The Russian fortress of Kinburn on the Black Sea was bombarded from the sea by an Anglo-French squadron in October 1855 during the Crimean War. The success of French iron-plated wooden, screw, floating gun platforms off Kinburn encouraged the French to build the ironclad *Gloire* which was laid down in 1858. **RIGHT**

VESSELS DESTROYED AT NORFOLK, 1861, BY THE UNITED STATES COAST SURVEY Secession by the Southern states led to the scuttling of federal warships off the naval yard at Norfolk. This subsequently became the centre of naval operations. **OPPOSITE**

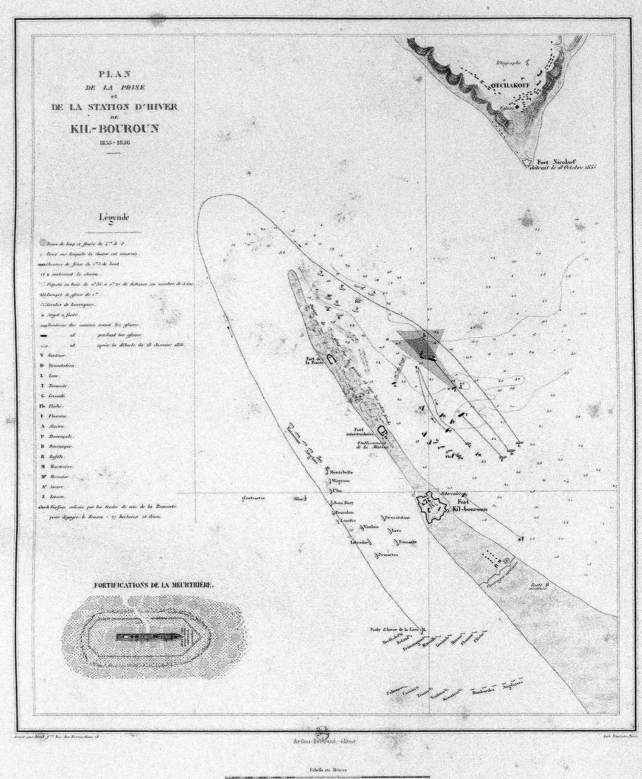

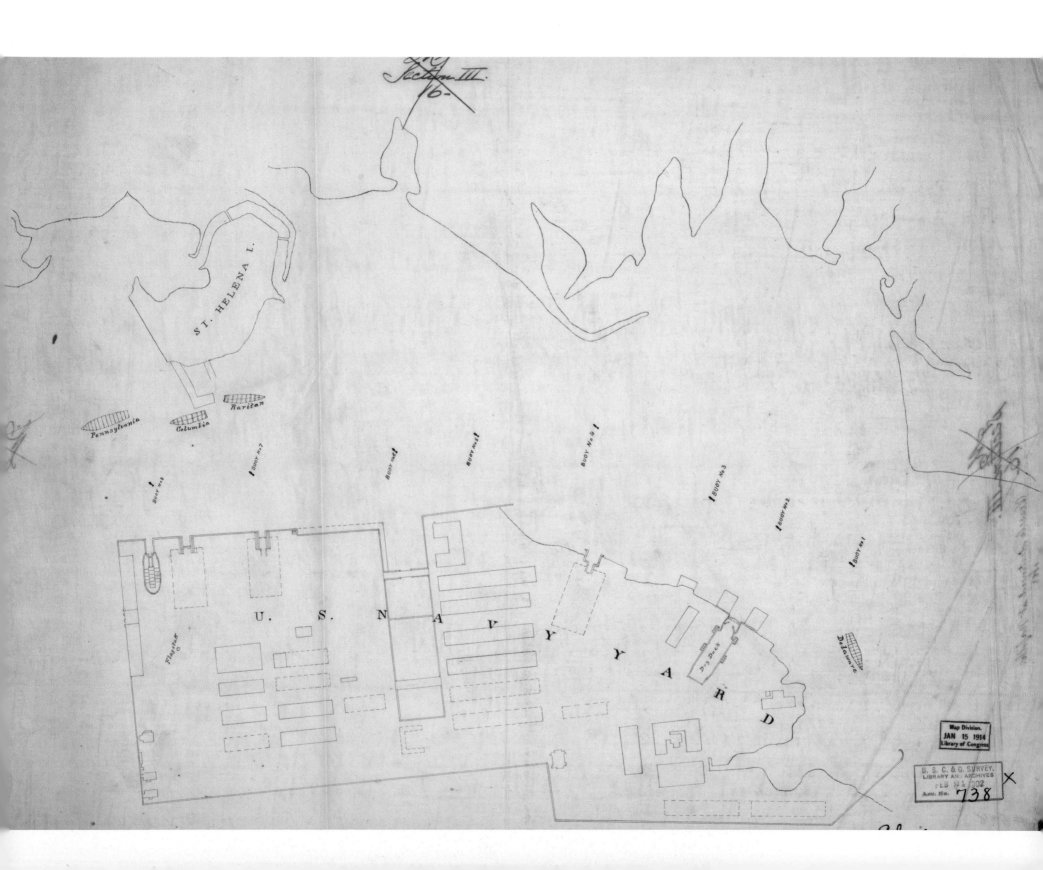

MAPPING NAVAL WARFARE

SEVERAL ROUTES PROPOSED FOR THE PASSAGE OF GUNBOATS TO THE LAKES, 1862 The routes to the Great Lakes included the Erie and Oswego Canal, and the Illinois River and Chicago Canal. Tensions between Britain and the Union increased from late 1861, with Britain fearing Union attacks on Canada, including on the Great Lakes. British concern rose as the Union discussed enlarging canals to enable warships to move to the Great Lakes, although ultimately Congress decided that it was more appropriate to build up its naval facilities on the lakes. In November 1861, Captain William Noble of the Royal Engineers emphasised an American threat to the St Lawrence River and the Welland Canal on the Niagara peninsula, and called for a British naval squadron on Lake Huron. In November 1864, Lieutenant Colonel William Jervois pressed for iron-plated vessels in order to protect the St Lawrence. Sweet's map shows monitors in the Atlantic. RIGHT

MAP
SHEWING THE SEVERAL ROUTES
Proposed for the
PASSAGE OF GUN-BOATS
TO THE LAKES.

VIA

Erie and Oswego   Canal

Champlain   "

Illinois River and Chicago   "

Wisconsin , , Green Bay

PREPARED BY

S. H. SWEET

DEP. STATE ENGR AND SURVEYOR

1862.

———— Canals

Distance by Sea Route NEW YORK to LIVERPOOL 3023 Miles
Distance by Lakes & Erie Canal from Str MACKINAW to N. YORK 1087 M.

MAPPING NAVAL WARFARE

FARRAGUT'S ATTACK ON THE MISSISSIPPI FORTS, 1862, BY ROBERT KNOX SNEDON This map shows the Confederate fortifications at Fort Jackson and Fort St Philip and the Union fleet under Farragut. To capture New Orleans, the largest city and principal port in the Confederacy, Farragut overcame the Confederate warships (the massive CSS *Louisiana* could not move for want of her engines, while the CSS *Manassas* only mounted one thirty-two-pounder) and bypassed the two forts at night, but only after the river was freed of obstacles. Off Manila in 1898, Dewey employed the technique he had observed when taking part in Farragut's attack: of passing heavily fortified shore positions at night. Farragut's success had not been matched by the British in 1815. The map included the longest range of fire from the forts.

RIGHT

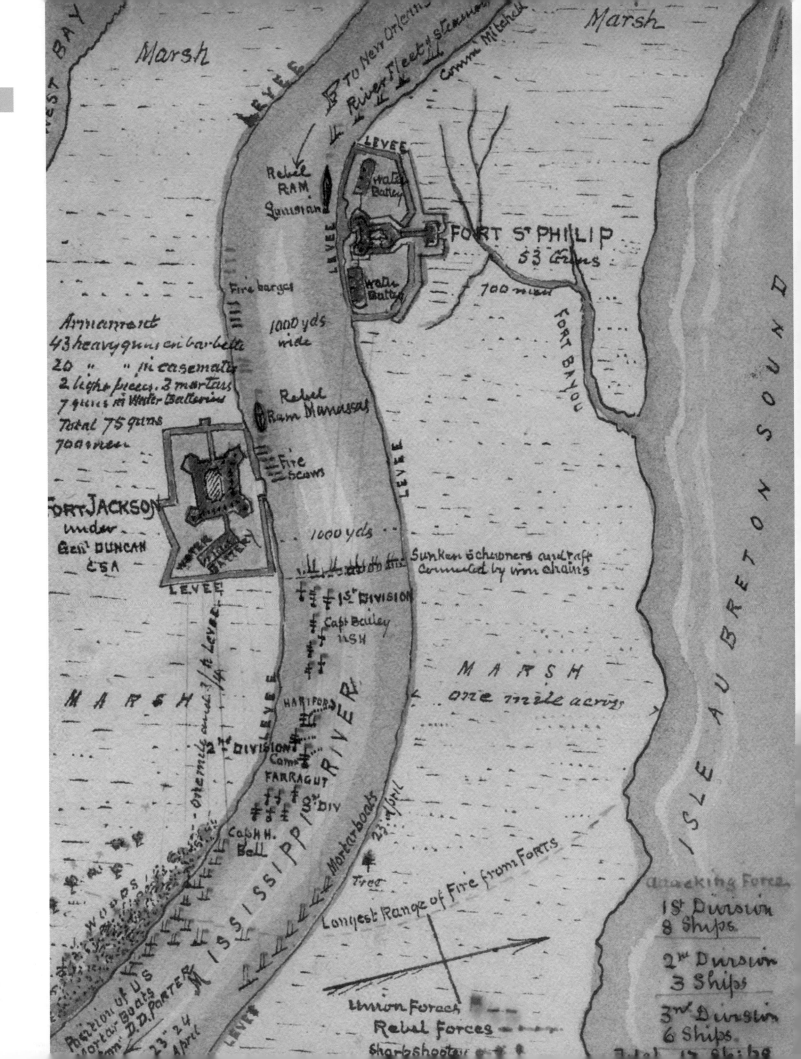

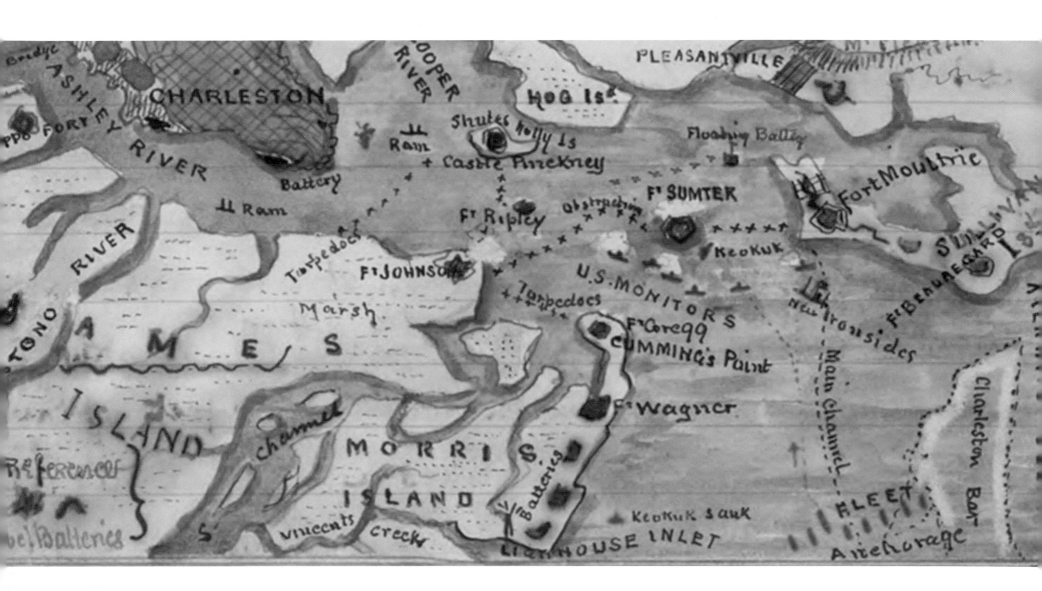

PLAN OF THE UNION ATTACK ON CHARLESTON, 1863, BY ROBERT KNOX SNEDON On 6 April 1863 Rear Admiral Samuel Francis du Pont commanded a powerful force of nine Union ironclads, but they were hampered by mines and exposed to fire from shore batteries. Benefiting from offshore islands, Charleston was protected by a network of defensive positions. One Union ironclad, the fixed-tower, thinly armoured USS *Keokuk*, was sunk, while three of the seven monitors were damaged enough to be sent for repairs, although they were all ready for action within a month. Charleston only surrendered in 1865, when threatened by General Sherman with a siege. In the American War of Independence, a British naval attack had failed in 1776 but a siege had succeeded in 1780. **ABOVE**

MAPPING NAVAL WARFARE

THE END OF A COMMERCE-RAIDER, 1864

This map accompanied Captain John
Winslow's detailed report of the sinking
on 19 June 1864 off Cherbourg of the
Confederate CSS *Alabama* by the Union's
USS *Kearsage*. It shows the position of
USS *Kearsage* when it received the first
broadside from the CSS *Alabama*, the
moves of the ships and the position of
the *Alabama* when it sank. In an attempt
to use steam warships to destroy enemy
commerce, the *Alabama* was a large (990
tons) and fast ship built at Birkenhead in
neutral Britain. Sailing from the Mersey
on 15 May 1862 the *Alabama* collected its
armament outside Britain in the Azores,
but there was little doubt of its intention,
and the ship captured between sixty-nine
and seventy-one merchantmen (sources
vary) before being sunk, an event seen
from shore and later recorded on canvas
by Edouard Manet. The episode came
close to leading to conflict between the
Union and Britain, and Britain paid £3.2
million in damages after the war. RIGHT

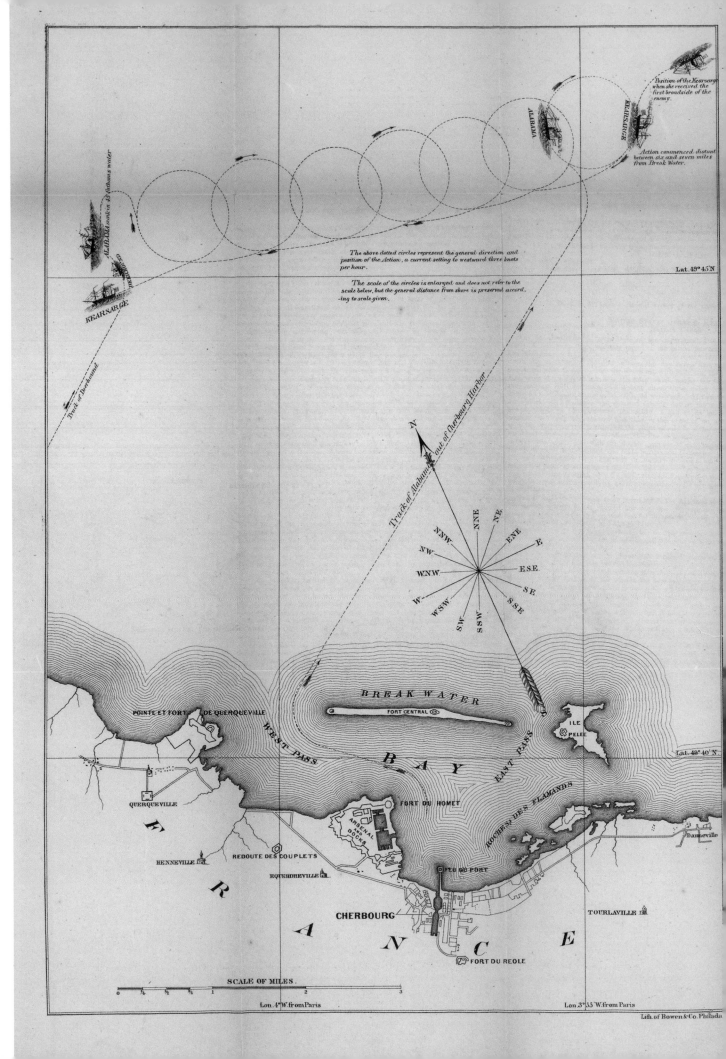

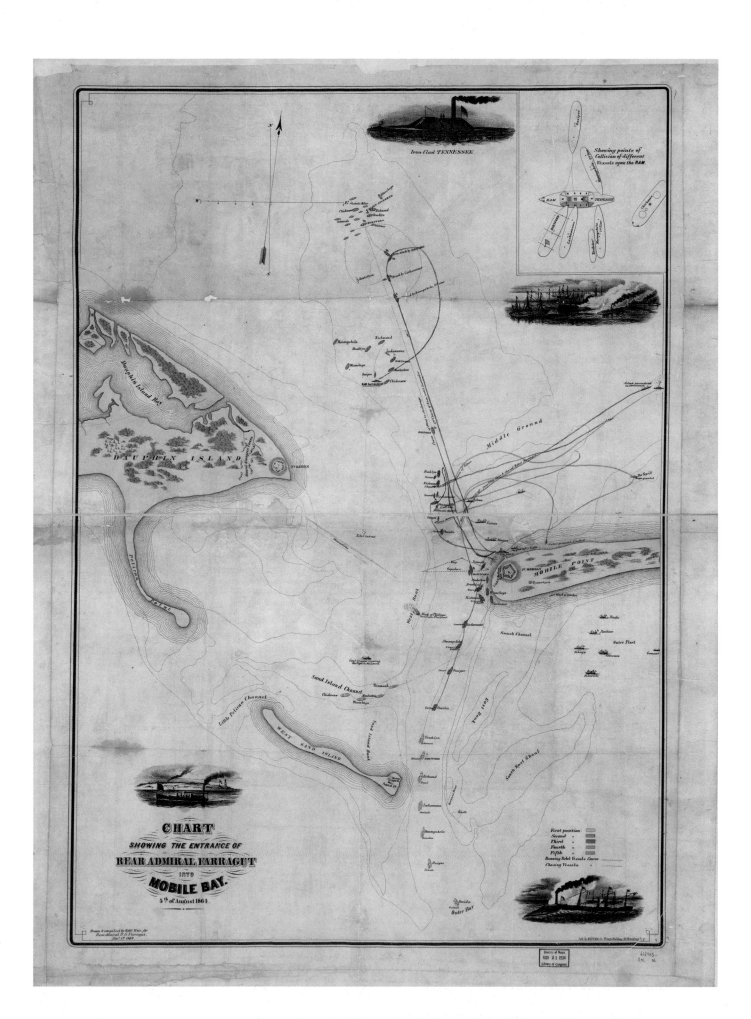

FARRAGUT AND MOBILE, 1864, DRAWN AND COMPILED BY ROBERT WEIR FOR REAR ADMIRAL FARRAGUT, 1 NOVEMBER 1864 The Union fleet successfully fought its way into Mobile Bay on 5 August 1864, despite mines which claimed one ironclad and could have claimed more had they worked better. This map offers much detail as well as visual support, including five positions of the warships and their route, obstructions, channels, banks and shoals. The diagram shows points of attack on the Confederate ironclad CSS *Tennessee*, while the views show this ship, a Union warship colliding with it, a ship steaming past Mobile Point and a general view of the battle. In 1815, after the close of the War of 1812, British forces had been preparing to attack the city having captured Fort Bowyer which protected the Bay. LEFT

MAPPING NAVAL WARFARE

ATTACK ON FORT FISHER, 1865, BY ROBERT KNOX SNEDON After an attack in December 1864 had failed, General Alfred Terry used a force of fifty-eight warships, the largest hitherto assembled in the war, to bombard Fort Fisher in North Carolina on 15 January 1865, before landing 8,000 troops. The capture closed the port of Wilmington, North Carolina. RIGHT

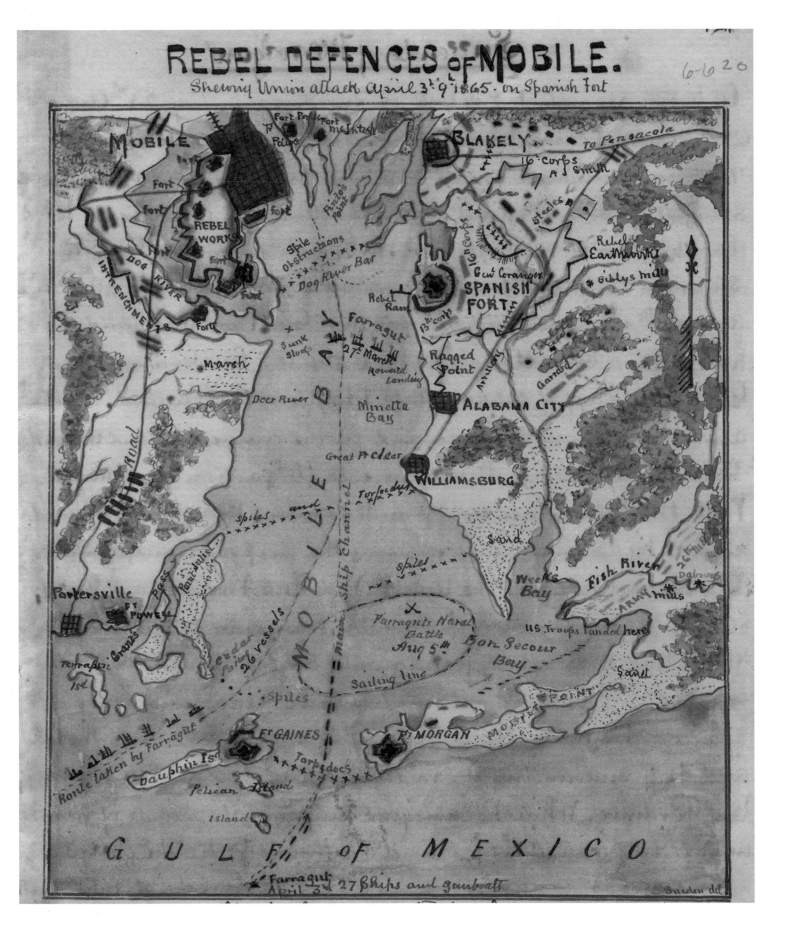

REBEL DEFENCES OF MOBILE, SHOWING THE UNION ATTACK OF 3–9 APRIL, 1865 BY ROBERT KNOX SNEDON The map shows the route taken by the Union squadron under David Farragut on 5 August 1864 when he successfully fought his way into Mobile Bay as well as in 1865 when his fleet provided support to forces under Frederick Steele and E. R. S. Camby that captured Spanish Fort and Blakely across the bay from Mobile, before entering that city on 18 April. The map noted fixed obstructions: spikes and torpedoes (naval mines). The latter claimed one ironclad on 5 August 1864 and could have claimed more had they worked better. **LEFT**

BATTLE OF MANILA, 1898, *FROM OUR COUNTRY IN WAR AND RELATIONS WITH ALL NATIONS* **BY MURAT HALSTEAD** In Manila Bay, the American Asiatic Squadron under the well-prepared Commodore George Dewey destroyed seven Spanish ships, essentially colonial gunboats, and silenced the shore batteries at the cost of eight Americans wounded on 1 May 1898. Having had most of his ships destroyed, Rear-Admiral Patricio Montojo surrendered. The absence of American bases in the region made this achievement especially impressive, as did the lack of reliable intelligence. The Spaniards had fewer large guns on their warships, but benefited from the shore defences, only for Dewey to employ a technique he had observed when taking part in Farragut's attack on New Orleans in 1862: passing heavily-fortified shore positions at night. Dewey's victory made naval success appear easy, but in practice it was hard-going because the gunnery was so uncertain. Only a small percentage of the American shells found their targets. Moreover, without troops, Dewey was unable to capture Manila and had to wait for the army's arrival. Fortified by the Americans, Corregidor, referred to as the 'Gibraltar of the East' fell to the Japanese in May 1942 after being exposed to heavy artillery fire and damaging air raids, and having been invaded at night. RIGHT

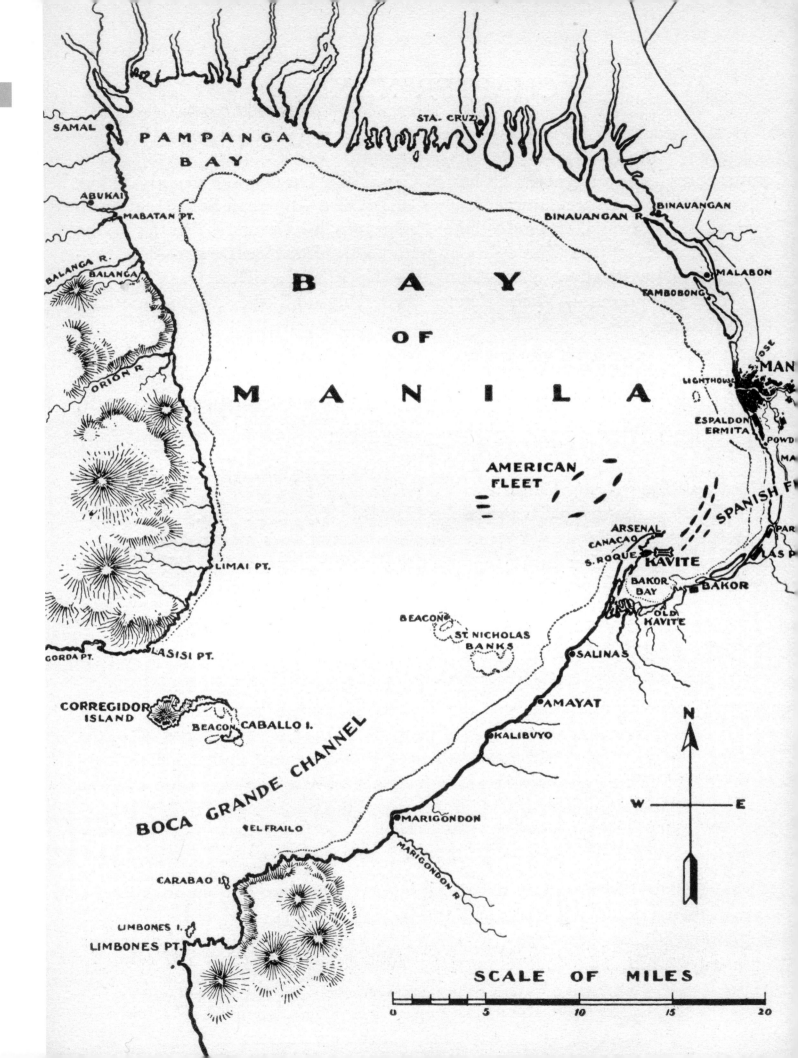

The War

SINCE we last wrote the war has made considerable progress, though not, perhaps, altogether in the direction that was expected, The invasion of Cuba is at length an accomplished fact, and there, at least, is nothing to cause surprise. The troops, so long delayed in the weary heat and dust of Tampa, received the intelligence that at length they were to embark with unbounded delight, and the vast convoy of transports put to sea with commendable celerity, and arrived off Santiago as we reported. The landing of marines at Guantanamo was either a feint or a misunderstanding, and the actual disembarkation took place at Baiquiri, about fifteen miles east of Santiago, where there is an iron pier. The operation began at 9.45 a.m. on Wednesday of last week, and was conducted with perfect success. The mobility conferred upon the Americans by their sea-base made it impossible for the Spaniards to resist the landing, and they wisely made no attempt to do so. Moreover, a body of insurgents, estimated at 5,000, had been watching the coast between Santiago and Guantanamo during the night. The hills behind Baiquiri, where, it was thought, Spaniards might be in force, were shelled by the *New Orleans, Machias, Detroit, Suwanee,* and *Wasp,* and boats carrying parties of the 8th, 10th, 12th, and 13th Infantry soon put off to the shore. There was a natural eagerness among the men to be the first to set foot on Cuban soil, the men scrambling over one another in their eagerness, and the landing was announced by loud cheers. The honour fell to the 8th Infantry. The disembarkation of troops continued throughout the day, and the whole force was not ashore until one o'clock on Thursday morning.

Each string of boats was greeted by the sounds of "Yankee Doodle," and the men were marched into quarters. At night General Lawton threw out detachments to the west and north, and the troops slept in deserted houses and in tents pitched near the village. General Linares, in command of about 1,200 Spaniards, abandoned Juragua upon the landing of the Americans, and retired in the direction of Sevilla, about five miles to the north-west, with the purpose of operating upon the flank of the hostile advance. General Lawton's brigade, consisting of the 1st and 22nd Infantry, the 1st Massachusetts Volunteers, and detachments of the 4th and 5th Cavalry, advanced by forced marches, suffering greatly from the intense heat, and at noon on Thursday occupied Juragua. There was some skirmishing, but no casualties occurred. Roosevelt's Rough Riders, the now famous body of society and college men, cowboys and others of bone and muscle, with other troops, hurried forward, but General Shafter, with the main body, did not reach Juragua until darkness had set in. The distress of the men during the forced march was very great, and for miles along the road from the landing place discarded uniforms and blankets were scattered, the soldiers declaring that their equipment was unsuited to the climate. Pack mules were not available, and great difficulty was found in forwarding supplies, while to bring up the siege guns was impossible without a great deal of labour requiring time.

The Spanish troops at Santiago number probably 20,000 men, and General Pando has advanced from Holguin with 10,000 more. Garcia was to have interposed, but it would appear that his force of 3,000 was considered inadequate, and he has since joined General Shafter. The combined force numbers about 20,000, but will soon be reinforced, and there are other insurgent bodies, who, however, cannot be depended on. The Spaniards have also troops to the north-east, with which they are said to threaten the American rear. But the real fighting must be close to Santiago, where fierce resistance is expected, and the great attack will not be made until further American forces can be brought to the front. From the positions reached by the Americans Santiago is visible, and it is seen that every height is surmounted by a blockhouse, while spies report that

trenches have been dug all round the city, with lines of stretched wire in front of them, all portending a desperate resistance. Admiral Cervera has reported the situation to be critical, and he has landed men from his vessels, but the guns of his vessels should be a powerful factor in the defence.

The first fighting of serious character occurred on Friday of last week. At daybreak General Young, with a body of regulars and Roosevelt's Rough Riders, under command of Colonel Wood, left Juragua, the purpose being to dislodge General Linares from the position he had assumed on the flank. The Rough Riders, who advanced dismounted over the wooded hill, were to take the Spaniards in front, while the regulars, taking a trail at the foot of the hill range, were to operate on the flank. In the event

MAP OF SANTIAGO HARBOUR
Showing the position of the sunken collier *Merrimac* and of the vessels of Admiral Cervera's squadron after the bombardment of May 31, as reported to Admiral Sampson by Cuban spies

of steep hills, the trail being so narrow that the men had often to march in single file. Each carried 200 rounds of ammunition and the heavy camping equipment, but, as the morning grew hotter, blankets, tent rolls, and empty canteens were thrown into the prickly cactus which bordered the way. When the men had marched about four miles the cuckoo-call of the Spaniards was heard, and the hilarious sounds of the march were sunk to a whisper. Halts were frequent, and great precautions were now taken, but it was impossible to see many yards. At about eight o'clock in the morning a place was reached where the track opened out, and there the dead body of a Cuban was discovered. The men loaded their carbines just in time, for sharp shots began to sound from the thicket and Mauser bullets to cut chips from the wood overhead. An ambush had been reached, and the firing of the Spaniards seems to have been excellent. Sergeant Hamilton Fish, a gentleman well known in American society, and grandson of a former Secretary of State of the same name, fell dead, shot through the heart.

Colonel Wood then advanced his men over the open ground, while Lieutenant-Colonel Roosevelt led a party through the thicket. Captain Capron, who was using his revolver effectively, fell mortally wounded as he urged on his men, saying, "Don't mind me, boys; go on with the fight." There was some confusion, but the men quickly rallied, and, after about fifteen minutes' heavy firing, the

two engagements occurred, the brunt falling upon the Rough Riders, who were very severely handled. Their advance was over a series

New Batteries · *Porter* · Entrance to Harbour · Old Fort · *Mayflower*

U.S. SHIPS GUARDING THE ENTRANCE TO SANTIAGO HARBOUR

SANTIAGO DE CUBA, 1898 Photography adds its account. This illustration for *The Graphic*, on 2 July 1898, shows American warships guarding the entrance to Santiago de Cuba, and provides a map of the harbour, the two offering very different views. Complete American victory was obtained on 3 July over the Spanish fleet off Santiago under Admiral Pascual Cervera, with the Spanish warships (four modern cruisers and two destroyers) sunk as and after they left the protected anchorage in an easy, but badly-commanded, American victory over warships in poor condition. This victory gave vital leeway in Cuba to the inadequately trained American army and encouraged the Spanish commander in Santiago to surrender. The battle demonstrated the ability of warships to deliver a decisive victory. LEFT

THE NAVY LEAGUE, MAP OF THE BRITISH EMPIRE,

1932 Founded in 1894, the Navy League was designed to orchestrate public pressure for naval strength. The map captures the role of maritime routes in maintaining the British empire and crucially its global trading system. The depiction of the evolution of the Royal Navy by means of ship types underlined the significance of naval strength. In 1932, however, there was already the challenge of relative naval weakness as a result of the growth of the American and Japanese navies. Britain was under pressure from economic problems and harsh fiscal circumstances, and was affected by the restrictions on naval strength agreed by the Washington (1922) and London (1930) naval treaties. Britain did not lay down any new battleships between January 1923 and December 1936, and in 1930 agreed to a limit of seventy cruisers. Naval expenditure was cut seriously in 1928–34. Nevertheless, despite a growing obsolescence, the Royal Navy was well trained and supported by an unrivalled number of bases. From 1936, helped by an increase in the navy estimates, the Admiralty was free to pursue ambitious policies. Many carriers, battleships, cruisers and destroyers were laid down, including the battleships of the 35,000 ton King George V class. **OPPOSITE**

Naval surveying and charting rose greatly in importance and scale during the First World War (1914–18). The need for charts reflected a variety of factors. At the outset, there were naval operations around the world as the Allies attacked German colonies and sought to track down German surface raiders (that is, warships that were not submarines). The first task involved amphibious operations that required a knowledge of inshore waters that, hitherto, had generally been poorly charted, for example in German East Africa (now Tanzania, Burundi and Rwanda) or in Germany's far-flung Pacific possessions. The tracking down of surface raiders was less problematic in terms of maps, but much of the campaigning against them took place in poorly charted waters.

Thereafter, there were fresh challenges. The First World War took far longer than had been anticipated and involved tasks that had not been expected at its outset. The entry of the Ottoman Empire (Turkey) into the war on the German side led to British power projection in part to prevent Turkish attacks on the British empire, but also in the search of an alternative to focusing solely on war with Germany on the Western Front in France and Belgium. British expeditionary forces were sent against Iraq (part of the Turkish empire), and an attempt was made in 1915 to knock Turkey out of the war by sending a fleet through the Dardanelles and against Constantinople (Istanbul). This project fell foul of Turkish shore batteries and minefields, both of which dramatised the need for a charting of inshore waters that included the depiction of defensive positions on land and sea. The Allied (British and French) warships were unable to knock out the shore batteries, which themselves served to protect the minefields. The Turks had benefited greatly from a pre-war British naval mission. The

Gallipoli expedition then developed into an ultimately unsuccessful attempt to seize the land along the Dardanelles, an attempt dependent on naval power.

The war did not see many amphibious operations thereafter, in large part because forces deployed by such means were regarded as vulnerable once landed, or as unlikely to be able to exploit initial advantages in the face of the movement of defensive forces. This happened to the Allied Salonika expedition in 1916, which after landing was hemmed in until the closing weeks of the war. However, the Germans used amphibious forces successfully in the Baltic Sea against Russia, especially in 1917.

NEW CHALLENGES

A far more novel challenge was posed by the German use of submarine warfare. Initially, the Germans had focused on their surface fleet, as had the British. Indeed, in 1916, looking for another Trafalgar, the Royal Navy steamed into battle against the Imperial German Navy in the North Sea off the Jutland Peninsula. But the outcomes of the two contests were starkly different: Trafalgar was a decisive victory in 1805 for Horatio Lord Nelson, while Jutland – the most mighty clash between battleships in history – was a virtual draw. Nevertheless, the significance of Jutland is much greater than the contrast with Trafalgar suggests.

The possibility of decisiveness in the age of steam had been revealed by the dramatic Japanese victory at Tsushima in 1905 in waters close to those where China and Japan compete today. The eleven major Russian battleships present at the battle were sunk or captured, as were four of the eight cruisers, with the damage inflicted by big Japanese twelve-inch guns. The Japanese only lost three torpedo boats. Just as with American naval victories over Spain in 1898, but more

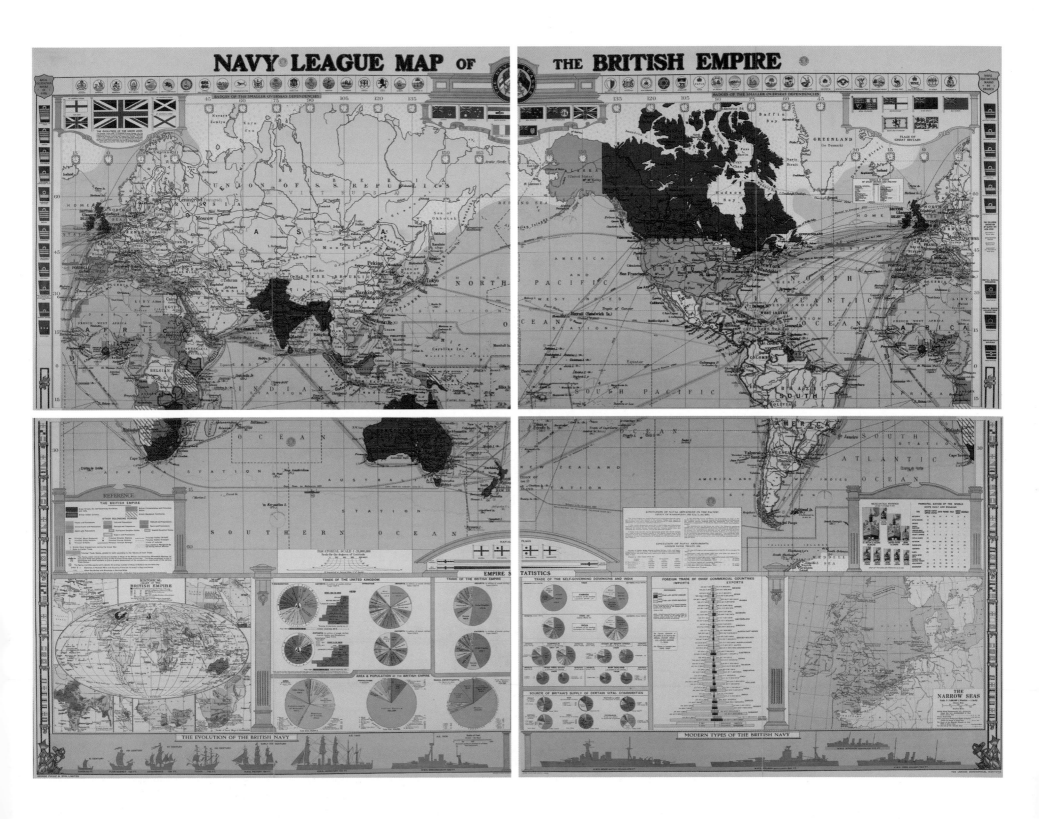

THE BATTLE OF TSUSHIMA, 1905, FROM *THE*
SYDNEY MORNING HERALD, **30 MAY 1905**
News of Japan's decisive naval victory
at Tsushima spread rapidly and helped
greatly to encourage concern about
Japanese power and ambitions, especially
in the USA, which was newly established
in the region (in the Philippines, and in
Australia). The 1902 Anglo-Japanese Naval
Treaty left Britain less anxious. Tsushima
prevented Russia from mounting naval
attacks on Japan and Japanese supply
routes to Korea, but it could not ensure
victory in the war. Moreover, the
subsequent cult of the Japanese
commander, Togo, proved misleading
when it was employed to argue that a
sweeping victory in battle would result in
success in war, an attitude that led Japan
to the totally misconceived attack on
the USA at Pearl Harbor in 1941.   RIGHT

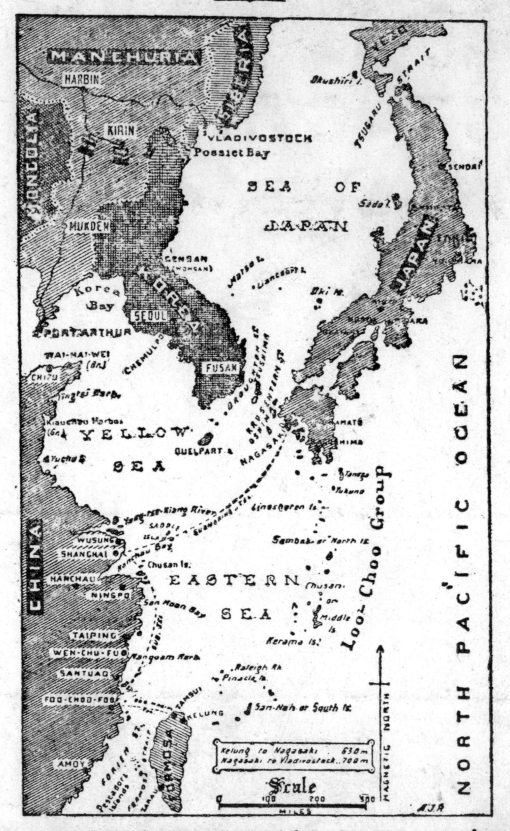

WHERE JAPAN WAS VICTORIOUS.

The above map enables the details of the historic naval battle, in the Straits of Korea between the fleets of Russia and Japan to be readily grasped. In the first place, it is advisable to note Formosa, at the bottom of the map. Bashi Channel (not marked) is south of Formosa, and on passing through its island-studded course the Baltic Fleet entered the open waters of the Pacific. That took place on Saturday, May 27. Admiral Roshdestvensky headed north, and anchored at the Saddle Islands (shown on the map), off Shanghai. There he coaled for the final stage of his 15,000 miles voyage, dividing his fleet, but taking all his principal fighting ships north-west to the Straits of Korea. Passing east of Quelpart Island, his fleet entered the straits, and was engaged in battle by the Japanese fleet, under Admiral Togo. Oshima Island has been mentioned in the cables in order to locate the battle. It will be seen that it is close to the coast of Japan—indeed, only 12 miles from the mainland. After sustaining appalling losses, including, it is believed, two out of four great new battleships, the Russian Admiral, in his flagship the Kniaz Suvaroff, steamed for Vladivostock with the remnant of his fleet. The map clearly shows one notable advantage possessed by Admiral Togo, viz., that to reach Vladivostock the Russian Fleet had to pass through narrow straits in order to gain the wide waters of the Sea of Japan. The result, according to "Reuter's," correspondent at Tokio, was the practical annihilation of the Russian Fleet. Incidentally we have shown the positions of the armies in Manchuria, the Japanese being represented by shaded signs to the southward. The particular point of interest at this hour is that the Russians have a strong force at Possiet Bay to resist an advance upon Vladivostock.

spectacularly, Tsushima appeared to vindicate the ideas of Alfred Thayer Mahan: an encounter would occur on the high seas, it could be a decisive battle and the result would then affect the fate of nations. Russia soon after accepted terms leaving Japan dominant in Manchuria and Korea. The battleships available to Britain and Germany in 1916 were far more powerful than those available to Russia and Japan in 1905, as the latter were pre-Dreadnoughts. Though worried about torpedo-boats, Russian and Japanese admirals, however, did not have to consider any threat from submarines.

The First World War exposed the difficulty of predicting developments and thus the limitations of much pre-war planning and speculation, as well as the problems of adapting in wartime to new technology, command and control systems, and doctrine. There were major issues of coordination, in the face of enemies whose capabilities and effectiveness were not known. The deficiencies of pre-war exercises contributed to the problem.

1914 and 1915 did not see any decisive naval campaigns in European waters. The clashes that occurred, notably the Anglo-German battles in the North Sea, at Heligoland Bight (1914), Texel (1914) and Dogger Bank (1915), were all small-scale. The most significant naval developments were the British success in 1914 in destroying German naval units outside Europe, particularly at the Battle of the Falkland Islands, where Graf von Spee's squadron was destroyed by British battlecruisers that had been rapidly sent there after a German defeat of an outgunned British squadron at Coronel off Chile. Moreover, amphibious operations captured the ports of German colonies.

German surface raiders had challenged Britain's supply system, that of a country that could not feed itself, an imperial economy that relied on trade, and a military system that required troop movements within the empire. That challenge was overcome. Allied success in blockading the North Sea, the English Channel and the Adriatic, and in capturing Germany's overseas colonies, ensured that, after the initial stages of the war and despite the use of submarines the range of German naval operations was smaller than those of American and French warships and privateers when attacking British trade in wars between 1775 and 1815. Germany could not match Britain's strength in surface shipping and, as had been predicted, suffered from Britain's position athwart German maritime routes to the Atlantic. This position was fundamental to the strategic situation.

Unlike in 1940, the situation was not transformed by German conquests. Whereas in 1940 the Germans conquered Denmark, Norway and France (and benefited from informal support from Spain in the shape of bases), making it far harder for the British to counter German naval operations, in 1914 the only ports conquered by the Germans were those in Belgium: Antwerp, Ostend and Zeebrugge. These ports provided the Germans with a stronger presence in the North Sea, but could not serve as fleet bases. In addition, Antwerp did not play a role comparable to that it had taken in British fears when at war with Napoleon. In practice, the German navy was still confined. This situation did not alter when Germany defeated Serbia (1915) and Russia (1917), nor when it inflicted a serious defeat on Italy (1917).

In 1915, the Germans responded to their failure to win on land in 1914, as they had anticipated doing on the Western Front, and to British naval dominance by stepping up submarine production and by launching, that February, unrestricted submarine warfare. This entailed attacking all shipping without restrictions and sinking without warning. RMS *Lusitania*, the largest liner on the transatlantic run, was sunk off Ireland by U-20 on 7 May 1915. Among the 1,192 passengers and crew killed were 128 Americans and there was savage criticism in America. In response, the Germans offered concessions over the unrestricted warfare, which was finally cancelled on 18 September, in order to avoid provoking American intervention.

Leaving aside the negative impact on neutrals, Germany was unprepared for unrestricted submarine warfare as it lacked sufficient submarines, trained crew or bases to mount an effective blockade of Britain. In early 1915, only twenty-nine submarines were available and, by the end of the year, only fifty-nine. In addition, submerged submarines were dependent on battery motors in order to move and these motors had to be recharged on the surface where the submarines were highly vulnerable. In a major assault, the Germans sunk 748,000 tons of British shipping in 1915, but Britain and its empire launched 1.3 million tons. Britain then had the largest shipbuilding industry in the world, and, at this time, this industry, and British ports, were not vulnerable to air attack, as they were from Germany in 1940–41.

JUTLAND AND ITS IMPLICATIONS

That was not the only strategic limitation of German submarine warfare. Submarines and mines appeared to be the means to snipe at the British naval advantage, rather than an effective response to it. As a result, reliance on the surface fleet still appeared necessary in 1916. The German plan then was to fall upon part of the British Grand Fleet with its entire High Seas Fleet. The plan had been tried in three other sorties earlier in 1916 that did not result in a battle, and was tried again in the Jutland operation.

Benefiting from superior Intelligence (due to signals interception) and analysis, the British did not fall for this plan. Nevertheless, despite having the larger fleet at Jutland, they failed to achieve the Trafalgar or, as it was then seen, sweeping victory hoped for by naval planners. This was an aspect of more general problems with learning to direct effectively the fleets of the period, not least issues of coordination and the use of Intelligence. At Jutland, the British suffered from problems with fire control, inadequate armour protection (notably on their battlecruisers), the unsafe handling of powder, poor signalling and inadequate training, for example, in destroyer torpedo attacks. German gunnery at Jutland was superior to that of the Royal Navy, partly because of better optics and better fusing of the shells.

Command decisions were also very important. Admiral Jellicoe's caution possibly denied the British the victory they might have obtained had the bolder Vice-Admiral Beatty, commander of the Battlecruiser Squadron (the advance British force), been in overall command. Beatty was regarded as more dynamic. However, as with the Duke of Wellington at Waterloo in 1815, Jellicoe only needed to avoid losing. It was famously remarked that he could have lost the war in an afternoon. He was worried that German torpedoes would sink pursuing British ships. Wellington was to win more than a defensive success at Waterloo: Napoleon not only failed but was routed. In contrast, Jellicoe did not achieve this outcome. However, circumstances were very different and the risks greater.

The British certainly lost more ships and men at Jutland than the Germans: fourteen ships, including three battlecruisers, and 6,097 men, compared with eleven ships, including one battlecruiser, and 2,551 men lost by the Germans. On 5 June 1916, four days after the end of the battle, Kaiser Wilhelm II, a

Pianta dimostrativa della Guerra Italo-Turca
LA BATTAGLIA NAVALE A KUNFIDAH

DEPOSITATA   (Riproduzione vietata su qualunque Scala)

PUBBLIC. PERIOD. - FOLIO N.º 7 - Gennaio 1912.

Casa Editrice E. BIAGIO GIARMOLEO - Via Plinio, 14 - MILANO - Telefono 76-58

ACTION BETWEEN THE *SYDNEY* AND THE *EMDEN*, 1914 Britain's supply system was that of a country that could not feed itself while its imperial military system required trans-oceanic troop movements. All of this was challenged by German surface raiders, but they were hunted down in the early months of the year. The *Emden* inflicted damage (and more) disruption on shipping in the Indian Ocean and shelled Madras (Chennai) in India. However, it was lost to the combination of naval fire from the Australian warship *Sydney* and a reef in the Cocos Islands on 9 November 1914. This was an important triumph for the Royal Australian Navy which had been established in 1911.

OPPOSITE

virulent Anglophobe, announced at the naval base of Wilhelmshaven: 'The English were beaten. The spell of Trafalgar has been broken.'

Nevertheless, the German fleet had been badly damaged in the big-gun exchange. Moreover, their confidence had been hit hard. Glimpses of the Grand Fleet had given German officers a view of Britain's formidable naval power. Thereafter in the war, the German High Seas Fleet sailed beyond the defensive minefields of the Heligoland Bight on only three occasions, the first on 18 August 1916. On each occasion, it took care to avoid conflict with the Grand Fleet.

In turn, precisely because the German High Seas Fleet had not been defeated, it continued to pose a threat as a fleet-in-being, rather as the French had done during the French Revolutionary and Napoleonic Wars. This threat acted as a restraint on British naval operations. The losses at Jutland made both Jellicoe and the Admiralty more cautious. British plans for bold large-scale operations, such as sorties into the Baltic Sea to help Russia, did not come to fruition. This was seen as significant as Russia was then under heavy pressure from German advances, and it was feared that Russia might be forced out of the war, as indeed happened in 1917–18.

Yet the British employed their fleet by deterring the Germans from acting and thus challenging the British blockade or use of the sea. This deterrence thwarted the German option of combining surface sorties with submarine ambushes in order to cause losses and reduce the British advantage in warship numbers (the method attempted by the Japanese against the Americans in the Midway Operation in 1942), and thereby acquire greater strategic capability. The advantage in capacity was supplemented by British superiority in the Intelligence war, especially

the use of Signals Intelligence organised by Room 40, which was important to British security as the location of German warships was generally known.

GERMAN STRATEGY AFTER JUTLAND

On 4 July 1916, recognising that Jutland had left the British still dominant in the North Sea, the German commander there, Vice-Admiral Reinhard Scheer, suggested to Wilhelm II, who took a close interest in naval operations, that Germany could only win at sea by means of using submarines. This, however, did not exhaust the German challenge. In particular, German destroyers and cruisers inflicted serious damage on two convoys trading with Scandinavia in October and December 1917. Further afield, German surface raiders were no longer a serious problem, although the potential for damage was revealed by SMS *Seeadler*, which sunk American cargo ships in the Pacific in 1917, before being shipwrecked near Tahiti, and SMS *Wolf*, an armed freighter despatched from Kiel in November 1916. In an epic journey before it returned home in February 1918, the latter sunk or mined twenty-nine Allied ships off South Africa, India, Sri Lanka, Australia and New Zealand, and created widespread alarm.

However, it was submarines that focused British concern after Jutland. In October 1916, Jellicoe observed that the submarine menace was getting worse. He attributed this to the greater size and range of submarines, and their increased use of the torpedo (as opposed to the gun) so that, to sink opponents, they did not need to come to the surface, where they were most vulnerable. The following month, Arthur Balfour, the First Lord of the Admiralty, wrote: 'The submarine has already profoundly modified naval tactics … it was a very evil day for this country when this engine of naval warfare was discovered.'

ACTION BETWEEN H.M.A.S SYDNEY
AND
S.M.S. EMDEN 9·11·14 SHEWING TRACKS
STEAMED

Sydney's track & positions shewn thus ——○——○——
Emden's track ·—·—·—·

NORTH KEELING Iᴰ

Landing Place

S.S. BURESK Collier

Shot fired to stop Collier.

S.M.S EMDEN 11·30 A.M. Aground and out of action.

H.M.A.S. SYDNEY 9·30 A.M. 9·11·14

10,500 yds. Emden opened fire 9·40 a.m.

S.M.S. EMDEN 9·30 A.M. 9·11·14

Items of Interest.

H.M.A.S. Sydney. Capt. J.C.T. Glossop, 5400 Tons, Speed 25 Kᵗˢ, Guns 8 - 6″

Shells fired 670, Casualties, Killed 4, Wounded 11. Distance steamed during action

69 Miles, S.M.S Emden, Capt. Von Muller, 3600 Tons, Speed 25 kᵗˢ Guns 10 - 4″·1.

Shells fired approx 1208, Killed 208, Wounded severely 45, slightly & unwounded 145.

DIRECTION Iᴰ

This is a true reproduction of the courses steamed by
H.M.A.S Sydney and S.M.S Emden while in action on the
9ᵗʰ November 1914.

Rear Admiral.

THE BATTLE OF THE FALKLAND ISLES, 1914 After defeating a weaker British force off Coronel on 1 November 1914, the German East Asiatic Squadron under Vice-Admiral Maximilian Graf von Spee was destroyed by a stronger British force off the British colony and naval base of the Falkland Isles on 8 December 1914. The autobiographical records of the sailors indicate a strong engagement. Henry Welch of HMS *Kent* reported on the sinking of SMS *Nürnberg*: 'We have avenged the *Monmouth*. I really believe it was in the *Nürnberg*'s power to have saved many in the *Monmouth*'s crew. Instead, she simply shelled her until the last part was visible above water. Noble work of which the German nation should feel proud. Thank God I am British.' These maps present different ways of recording and presenting the battle. OPPOSITE AND NEXT FOUR PAGES

The greater emphasis on submarines altered the nature of the war at sea, as submarine warfare did not offer the prospect of a decisive warfare in a climactic engagement. Instead, the submarine conflict ensured that war at sea became attritional, indeed much more so than that on land. The 'battle' now lasted months, if not years. Strategy was thereby transformed. There was a shift from a focus primarily on battleships, and before on ships of the line, to a more modern, industrial war, in which the sinking of merchantmen was of equal importance.

Jutland thus ensured a change in the content and tone of the war. Combined with the British blockade of Germany, the submarine conflict meant that the war was more clearly one between societies. There was an attempt to break the resolve of peoples by challenging not only economic strength, but also social, and indeed demographic, stability. The post-Jutland German air assault on London, launched in May 1917, was an aspect of this new focus, again putting civilians in the front line.

The challenge necessarily directed attention to the ability of governments to safeguard the Home Front. This ability became more important as the war continued without any end in sight. The absence of any diplomatic or military breakthrough suggested that the war would continue for a long time. Indeed, British hopes from the blockade were focused on victory in the early 1920s. This was an economic warfare in which the position of neutrals, such as the Netherlands, Norway, Sweden and, for long, the USA, was very important.

Linked to this, the shift of German naval strategy after Jutland led to American entry into the war. This entry was crucial due to America's military, economic and financial strength. The failure of the Jutland Operation as an attack on the British fleet (combined

with the inability to drive France from the war at Verdun, as the German army had anticipated by forcing a battle designed to cause heavy casualties, and the experience of the lengthy and damaging British attack in the Somme offensive, both also in 1916) led to the German determination to use submarines in order to force Britain from the war. There was a parallel to the German invasion of France via Belgium in 1914, in that the strong risk that a major power would enter the war as a result, Britain in 1914 and the USA in 1917, was disregarded on the grounds that success could be obtained as a result of the attack. A misguided belief in the certainty of success characterised German policymaking during the war, as with the invasion of France in 1914. There was a mistaken assumption that the options of others could be predetermined.

THE EFFECT OF AMERICA'S ENTRY INTO THE WAR

In 1916–17, the Germans had had plentiful warnings of the likely American and British responses as a result of their earlier use of unrestricted submarine warfare, but there was an ideology of total war and a powerful Anglophobia in nationalist right-wing circles. On 31 January 1917, Germany announced, and, on 2 February resumed, unconditional submarine warfare. It was seen as the way to win rapidly. It was also assumed that the submarines would be able to impede the movement of American troops to Europe. Thus, after Jutland had failed to have strategic effect, the submarine was given the task.

Ironically, America's entry into the war itself increased the very importance of submarines to German capability as it further shaped the balance in surface warships against Germany. This was a balance that had not been significantly shifted at Jutland. America had the third largest navy in the world after

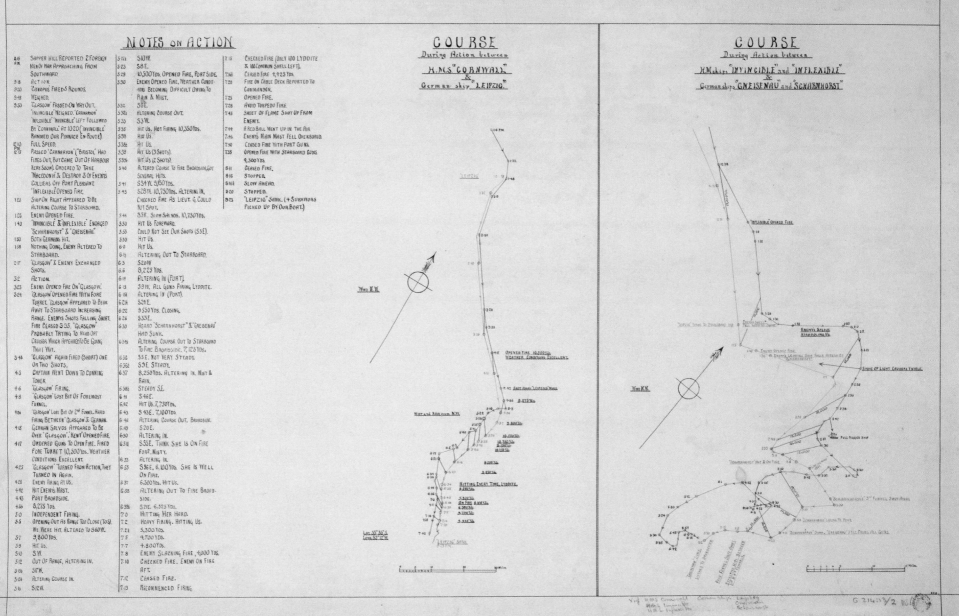

TRACK CHART of Battles off the FALKLAND ISLES 8th December 1914

Britain and Germany, and the Naval Act of 1916 had increased the American shipbuilding programme. In part due to the dominance of the army's needs in Germany, where there was no influence comparable to that of the British Admiralty, the Germans added fewer battlecruisers and, in particular, battleships to their fleet during the war than the British, despite having a large shipbuilding industry. Thus, linked to the results of Jutland, the Germans could not fall back

on a large-scale shipbuilding programme. Nor, more seriously, did they have the prospect of support from the warships of new allies that the British gained with the alliance of Italy (1915) and the USA (1917), the latter also providing valuable support in convoying merchantmen and troop transports. In addition, in late 1916, in accordance with a British request, four Japanese warships were sent to the Mediterranean. Based in Malta, they added to escort capacity as well

TRACK CHART OF BATTLES OFF THE FALKLAND ISLES ON 8 DECEMBER 1914 showing the course of action between HMS *Cornwall* and the German ship *Leipzig* and HMS *Invincible* and *Inflexible* and the German ships *Gneisenau* and *Scharnhorst*. ABOVE

MAPPING NAVAL WARFARE

A CHART SHOWING THE APPROACH OF THE GERMAN SHIPS *GNEISENAU* **AND** *NURNBERG* in an attempt to shell Port Stanley in the Battle of the Falkland Isles, 8 December 1914. **BELOW**

A CONCENTRIC CIRCLES CHART showing mooring and manoevring for fleet purposes during the Battle of the Falkland Isles in December 1914. **RIGHT**

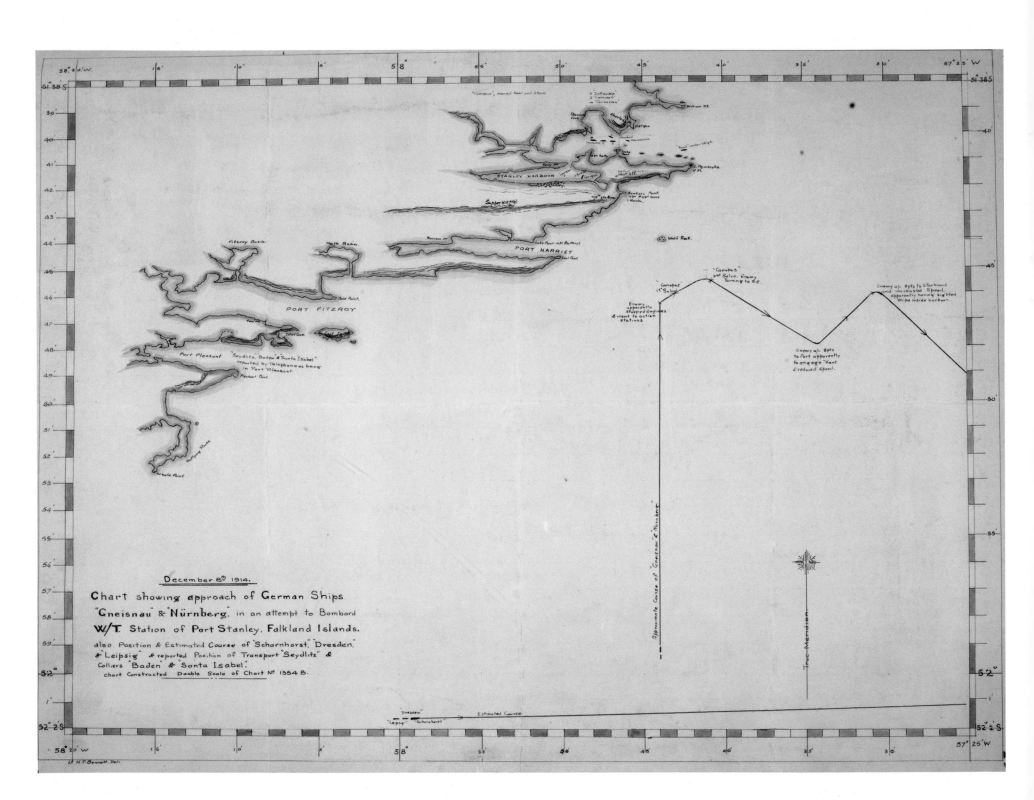

MOORING AND MANŒUVRING BOARD FOR FLEET PURPOSES.

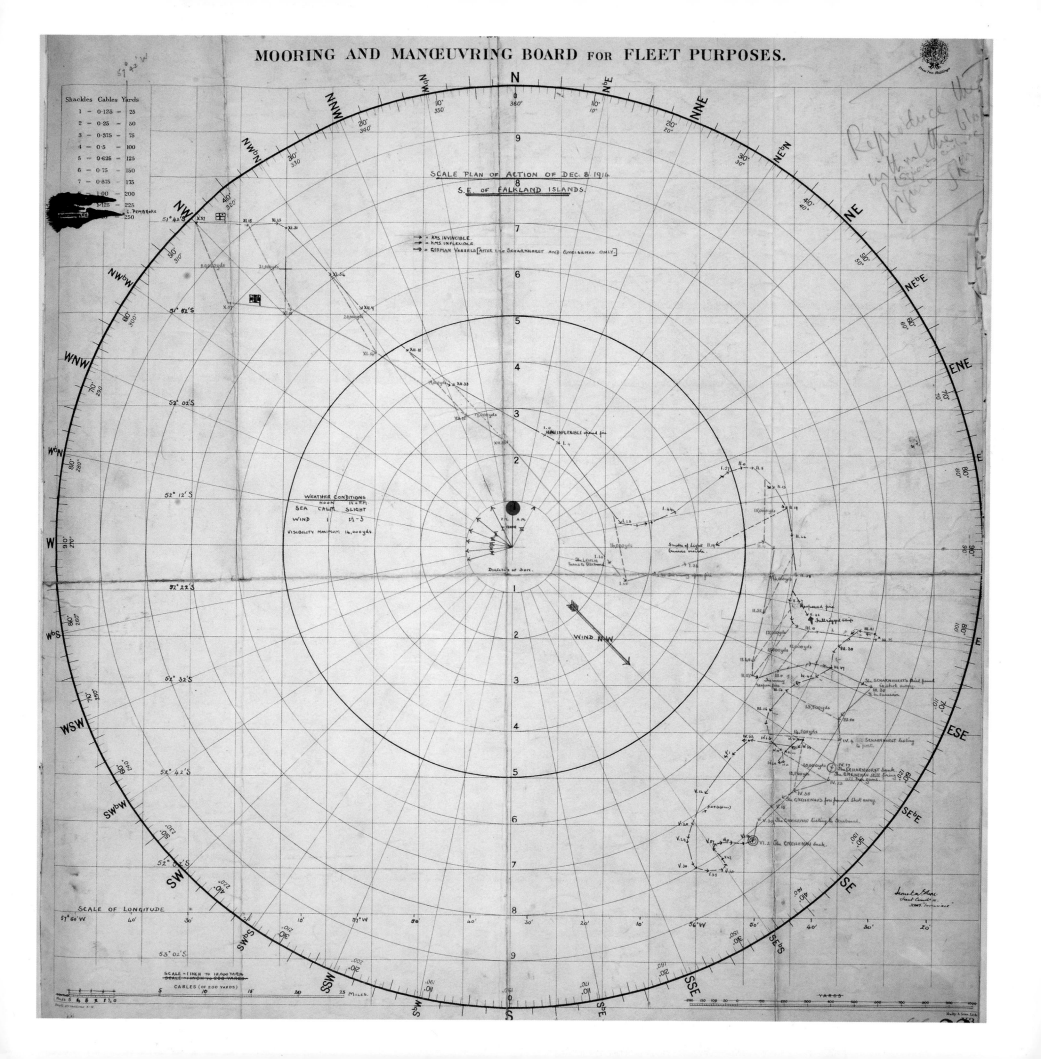

SCALE PLAN OF ACTION OF DEC. 8. 1914
S.E. OF FALKLAND ISLANDS.

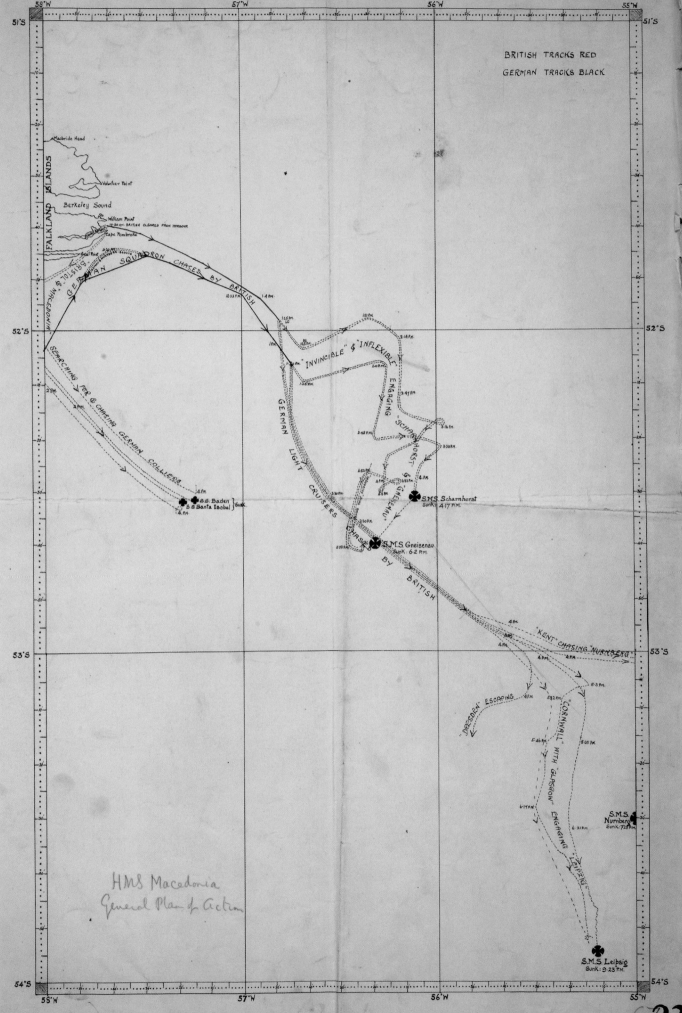

BRITISH TRACKS RED
GERMAN TRACKS BLACK

HMS Macedonia
General Plan of Action

as strengthening the Allied position in the equation of naval power. These additions more than nullified the success of German submarines in sinking Allied warships: in 1917, the British lost only one pre-Dreadnought battleship and one cruiser. Thus, the German failure in surface ship warfare in Jutland was matched by the more general arithmetic of surface ship strength. This arithmetic was a cause, means and product of the strategic defeat of the Germans at sea, and was matched by a failure at sea. The naval war was crucial to that on land, as it ensured the availability to move troops and provide supplies.

SYMBOLISM OF WAR

The war also had symbolic force. For the British, for example, the resolve shown by the battlecruisers at Jutland was regarded as heroic. William Lionel Wyllie, one of Britain's best maritime artists, produced works showing the battlecruisers HMS *Tiger, Princess Royal, Lion, Warrior* and *Defence* going into action and the loss of HMS *Invincible*, in which he had hoped to sail. Wyllie's work on Jutland, for which there was a German counterpart, represented a continuation and updating of his celebration of the nation's naval enterprise, as his works included a panorama of Trafalgar and a painting of HMS *Bellerophon* in Portsmouth with, in the background, HMS *Victory*, Nelson's Trafalgar flagship, which Wyllie had helped preserve for the nation.

The symbolism was also present in the link with royalty. George V's son, Prince George, served in the Royal Navy at Jutland. Later, as George VI, he was king during the Second World War, one in which the combination of the enmity of Germany, Italy and Japan, with the consequences of the German conquests of Norway and France, was to test the Royal Navy even more savagely than the earlier struggle.

NEW REQUIREMENTS FOR MAPS

Map-making had to keep pace with the war. In Britain, the Royal Naval Hydrographic Office, the staff of which increased to 367 in 1918, provided operational material for maps, but also charts for all merchantmen under the control of the Ministry of Shipping. This ensured that classified information could be made available about minefields, wrecks and buoyage. Correspondingly this information had to be denied to opponents. Naval operations, including the war with submarines, drove forward the need for maps, and ensured a new variety and complexity in what had to be depicted. In particular, with submarines, even the primitive and not very deep-diving submarines of the First World War, information on depth became far more important. Until the advent of submarines, any depth over 50 metres (including tides) in maps made no military sense, with the exception of minefields. The increased speed and range of warships were further issues.

So also was the reconnaissance information made available by aircraft and airships. In Bernard Partridge's cartoon 'Neptune's Ally', published in *Punch* on 25 May 1914, Winston Churchill, the first Lord of the Admiralty, was depicted as blowing aircraft and airships forward to aid the Royal Navy in protecting Britain from invasion.

Some of the additional information from this period was not useful until later in the twentieth century. This was particularly true of the extensive Arctic exploration of the 1890s and the first two decades of the century. Apart from Roald Amundsen achieving the first Northwest Passage in his ship *Gjøa* in 1903–06, the exploration revealed that the Arctic waters were deep and not, as earlier believed, shallow. This had major implications for submarine operations during the Cold War.

PLAN OF ACTION OFF THE FALKLANDS ISLANDS ON 8 DECEMBER 1914 The tracks of the British ships are show in red and those of the German ships in black. OPPOSITE

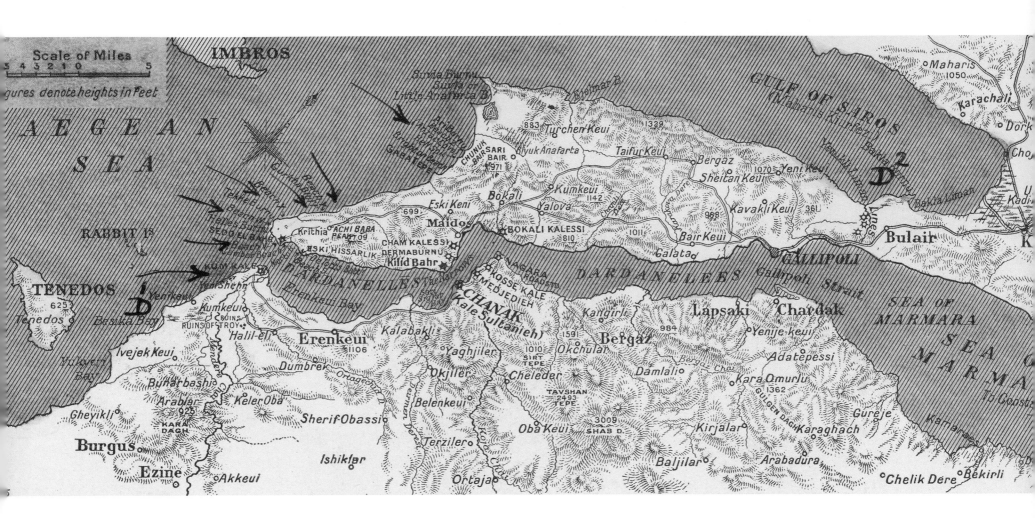

LITHOGRAPHIC MAP ANNOTATED BY HERBERT

HILLIER TO SHOW LANDINGS ON 25 APRIL 1915

This represented a continuation of the
pre-war concept of combined operations
as a means to support the bringing to
bear of naval power. However, the
campaign was launched against
strongly-defended positions and not, as
planned prewar, against poorly-defended
ones. A lack of appropriate equipment,
notably powered landing-craft, was a
major problem. **ABOVE**

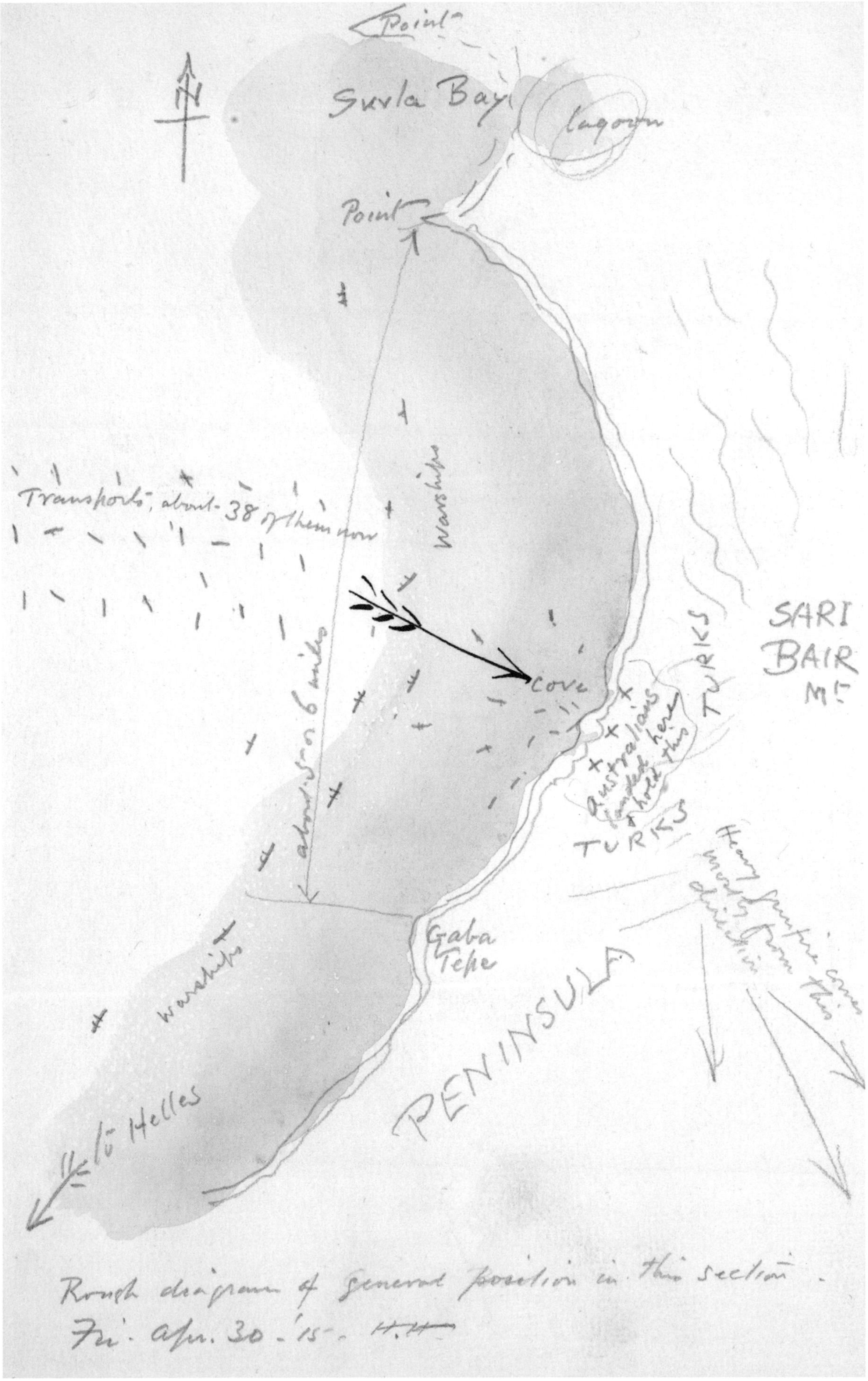

ANZAC LANDING ZONE, SKETCHED DIAGRAM BY HERBERT HILLIER, 30 APRIL 1915 The Allies managed to get their troops ashore, but were not able to overcome their opponents and achieve broader strategic objectives. **LEFT**

Point

Suvla Bay

lagoon

Point

Transports, about 38 of them now

Warships

about six 6 miles

Cove

Australians landed here hold this

SARI BAIR Mt

TURKS

TURKS

Heavy gunfire comes mostly from this direction

Gaba Tepe

warships

to Helles

PENINSULA

Rough diagram of general position in this section.
Fri. Apr. 30 - '15 - H.H.

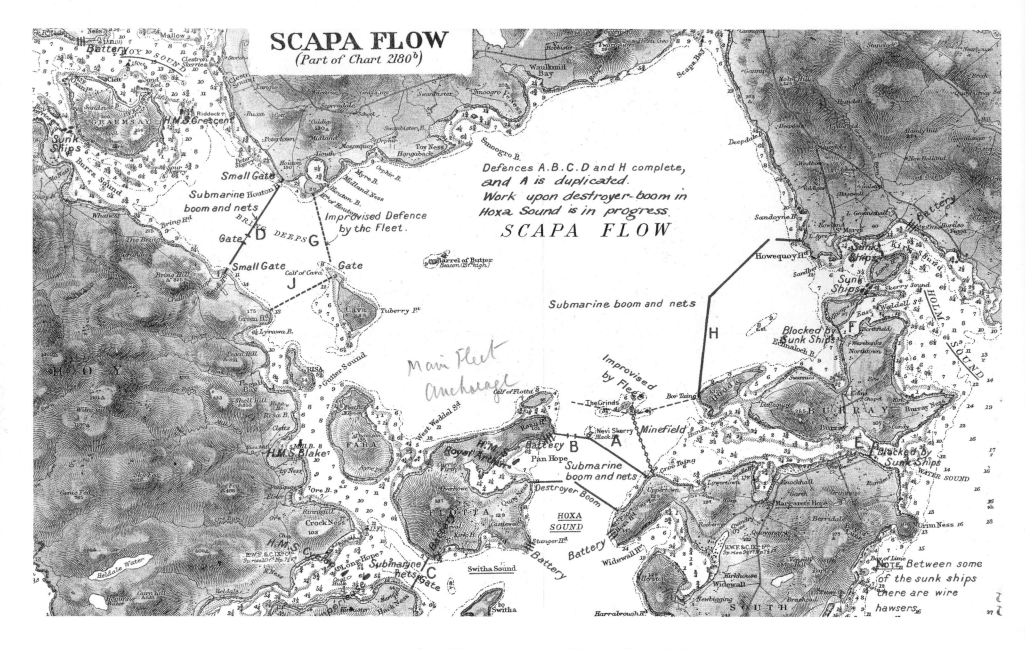

SCAPA FLOW
(Part of Chart 2180ᵇ)

Defences A.B.C.D and H complete, and A is duplicated. Work upon destroyer-boom in Hoxa Sound is in progress.

SCAPA FLOW

Main Fleet Anchorage

BRITISH NAVAL ANCHORAGE AT SCAPA FLOW, 1915
The German route to the Atlantic was threatened by the base of Scapa Flow in the Orkneys, which anchored the British naval position. The sinking of three British cruisers, all to the submarine SM *U-9*, on 22 September 1914, led the British Grand Fleet to withdraw to the north-west coast of Scotland. It did not return to Scapa Flow until 1915, when the base's defences had been strengthened. Scapa Flow continued to be a key operational asset, and an indication of the strategic weakness of submarines. ABOVE

DOGGER BANK, 1915 A German force of four battlecruisers and four light cruisers under Admiral Franz von Hipper put to sea in order to lay mines in the Firth of Forth, threatening the British base at Rosyth, and also to attack the British fishing boats on Dogger Bank in the North Sea, boats seen as an intelligence asset, but the interception of a German naval signal led to a loss of surprise. On the morning of 24 January 1915, five British battlecruisers under Vice-Admiral David Beatty engaged the Germans. However, in a stern chase, it proved inherently time-consuming to close, and this problem enabled the German ships to concentrate on his leading ship, the flagship *Lion*, which took serious damage. By contrast, confused signalling by Beatty, the fear of submarine attack, and a lack of initiative by subordinate commanders, ensured that the British force focused on the *Blücher*, a cruiser, which was sunk, only for the other German ships to escape. The British were affected by the contrast between the long range at which shells could be fired and their limited number of hits, while heavy smoke restricted the optical range-finding crucial to gunnery. RIGHT

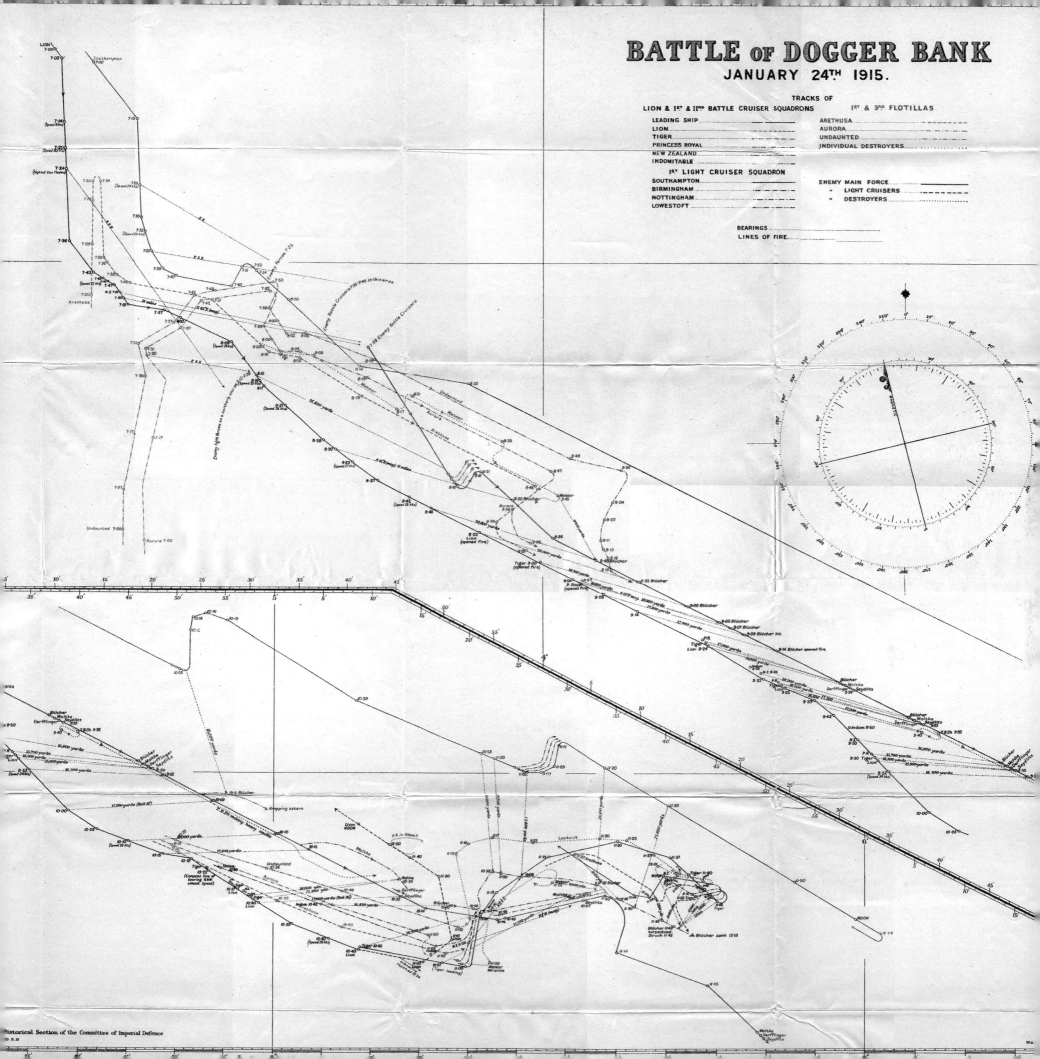

BATTLE of DOGGER BANK
JANUARY 24TH 1915.

TRACKS OF

LION & IST & IIND BATTLE CRUISER SQUADRONS	IST & 3RD FLOTILLAS
LEADING SHIP	ARETHUSA
LION	AURORA
TIGER	UNDAUNTED
PRINCESS ROYAL	INDIVIDUAL DESTROYERS
NEW ZEALAND	
INDOMITABLE	

IST LIGHT CRUISER SQUADRON

SOUTHAMPTON	ENEMY MAIN FORCE
BIRMINGHAM	" LIGHT CRUISERS
NOTTINGHAM	" DESTROYERS
LOWESTOFT	

BEARINGS
LINES OF FIRE

Historical Section of the Committee of Imperial Defence

ROBUR WAR-MAP, 1915 A bird's eye view of the Gallipoli operation issued for the Robur Tea Company of Melbourne, Australia. There are two maps on one sheet and they depict Allied landings as well as the range of naval guns while the relief is shown pictorially. A note advises that 'These maps (together with a set of flags) may be obtained by sending your name and address and four penny stamps to the Manager Robur Tea Company.' The inset map shows the strategic significance of the Dardanelles. To its supporters, this campaign appeared to be a viable alternative to the effort required in the damaging confrontation with the Germans on the Western Front, a means to bring decisive pressure on Turkey by exposing its capital, Constantinople (Istanbul) to bombardment, and a way to unblock the Balkans. RIGHT

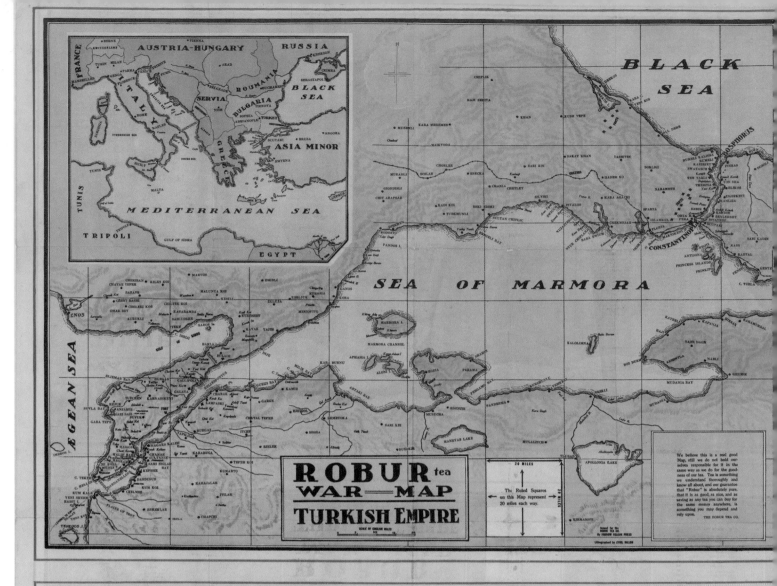

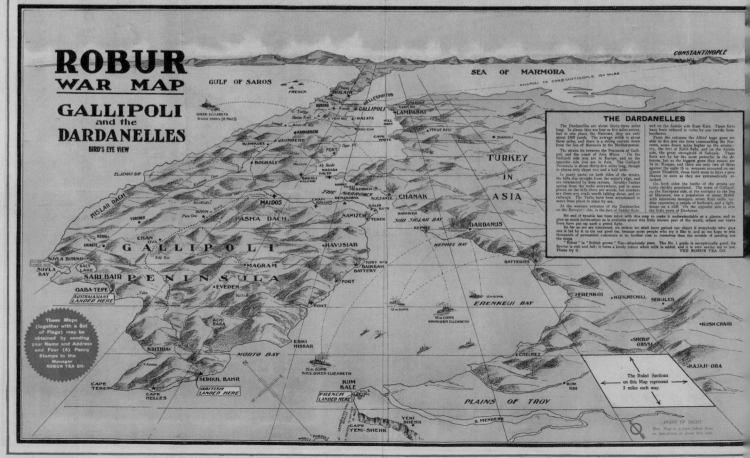

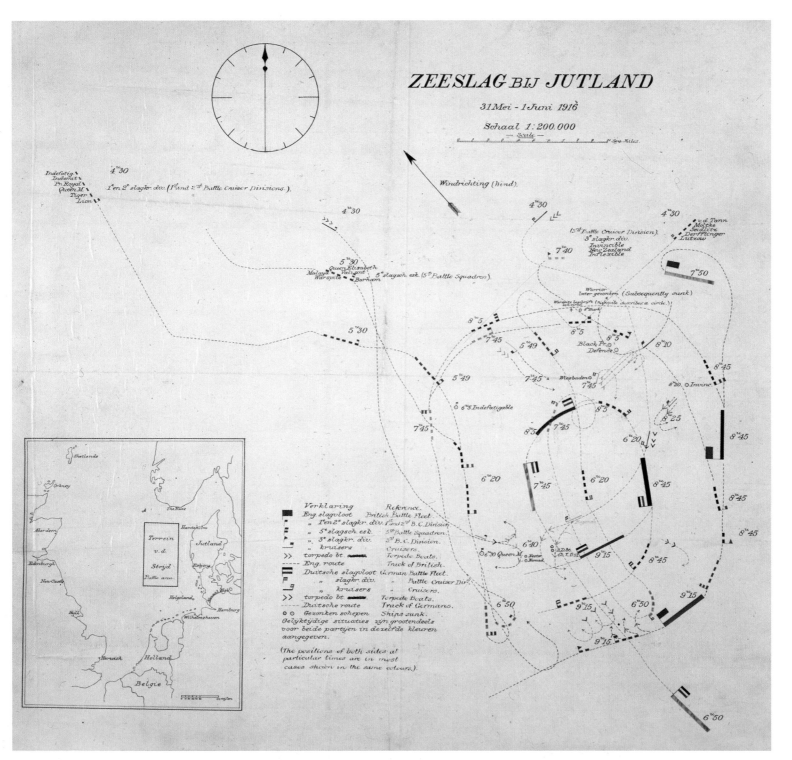

ZEESLAG BIJ JUTLAND

31 Mei - 1 Juni 1916

Schaal 1:200.000

— Scale —

Windrichting (Wind).

GERMAN MAP OF THE BATTLE OF JUTLAND, 1916

The Battle of Jutland, on 31 May–1 June 1916, was the most important battle in the age of battleships, and one between the two leading navies in the world. As aircraft and submarines did not take a significant role, this was a battle involving a relatively limited range of ships and weapon systems. Although destroyer torpedo attacks played a role, the key element was gunnery. This revealed flaws in the British battlecruisers, both the ships themselves and the handling of them, but the firepower and gunnery skill of the British battleships were much greater, and the Germans were fortunate to disengage. The battle confirmed Britain's strategic dominance. **LEFT**

MAPPING NAVAL WARFARE

THE BATTLE OF JUTLAND, 31 MAY – 1 JUNE 1916

The German plan was to turn upon part
of the British Grand Fleet with their
entire High Seas Fleet. The British did
not fall for this plan but, despite having
the larger fleet at Jutland, failed to
achieve the Trafalgar or, as it was then
seen, sweeping victory hoped for by
naval planners. Instead, the British
suffered from problems with fire control,
inadequate armour protection, notably
on their battlecruisers, the unsafe
handling of powder, poor signalling
and inadequate training. **RIGHT**

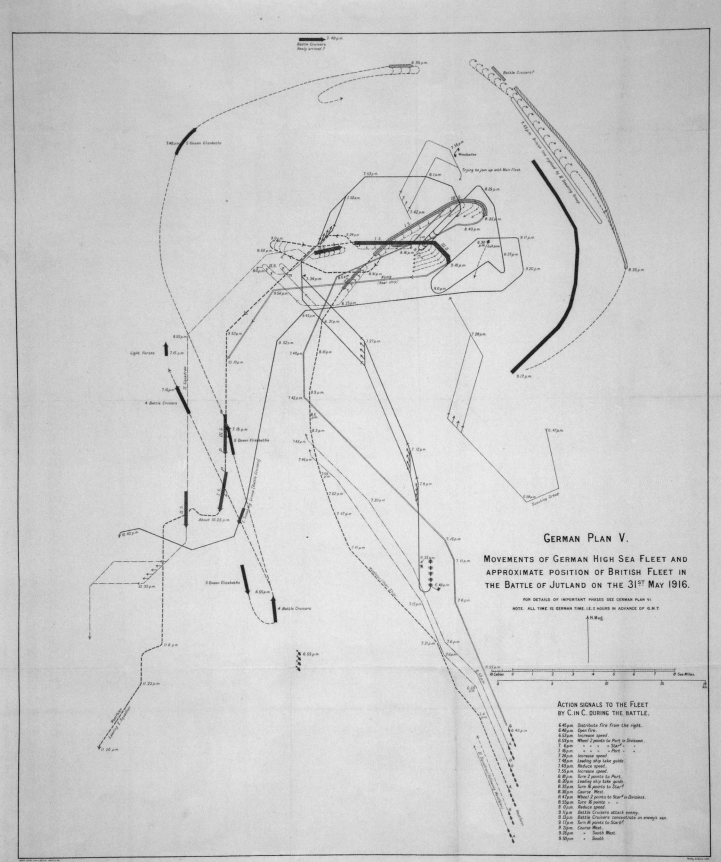

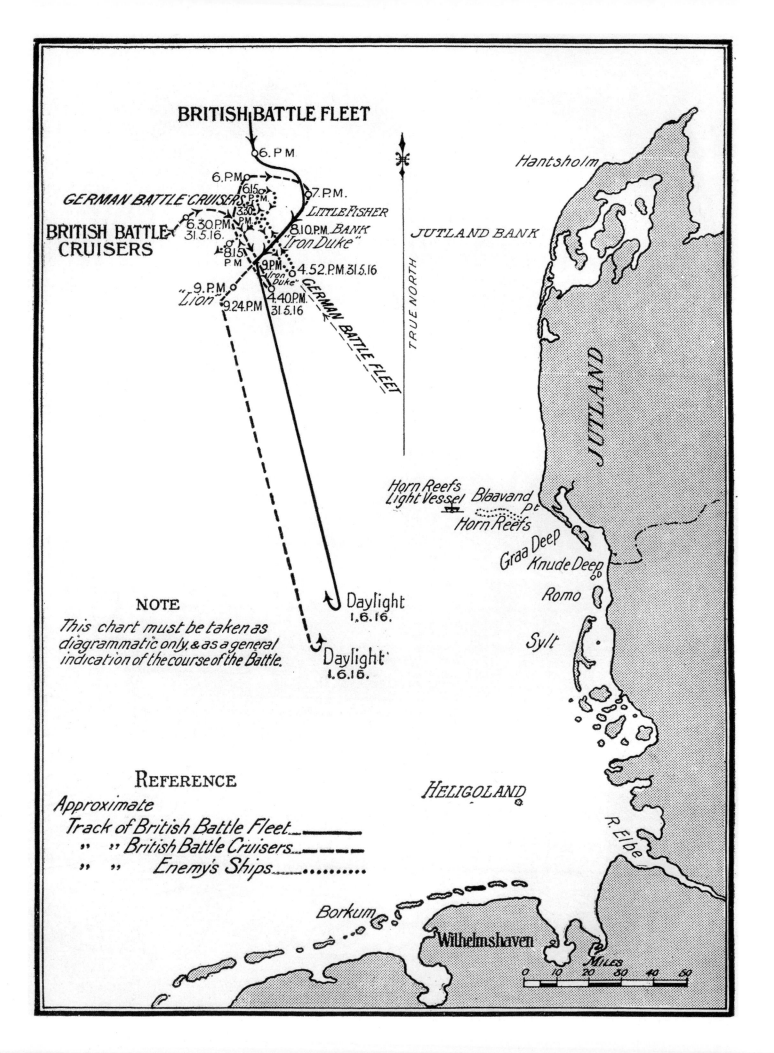

BRITISH BATTLE FLEET

6. P.M.

6. P.M.

GERMAN BATTLE CRUISERS

6.15 P.M.

3.30 P.M.

7. P.M.

LITTLE FISHER

BRITISH BATTLE CRUISERS

6.30 P.M.
31.5.16.

8.10 P.M. BANK

"Iron Duke"

8.15 P.M

9 P.M.

4.52 P.M. 31.5.16

"Iron Duke"

9.P.M.

"Lion"

9.24 P.M

4.40 P.M.
31.5.16

GERMAN BATTLE FLEET

TRUE NORTH

Hantsholm

JUTLAND BANK

JUTLAND

Horn Reefs
Light Vessel

Blaavand
pt

Horn Reefs

Graa Deep

Knude Deep

Romo

Sylt

Daylight
1.6.16.

NOTE

This chart must be taken as
diagrammatic only, & as a general
indication of the course of the Battle.

Daylight
1.6.16.

REFERENCE

Approximate
Track of British Battle Fleet
"   " British Battle Cruisers
"   "   Enemy's Ships

HELIGOLAND

R. Elbe

Borkum

Wilhelmshaven

MILES
0   10   20   30   40   50

THE BATTLE OF JUTLAND, 31 MAY – 1 JUNE 1916

Chart showing the course of the British
Fleet In the Battle of Jutland Bank during
World War I. From *The Year 1916
Illustrated.* LEFT

THE BATTLE OF JUTLAND, 31 MAY – 1 JUNE 1916

'Prepared in the Historical Section of the
Committee of Imperial Defence'. RIGHT

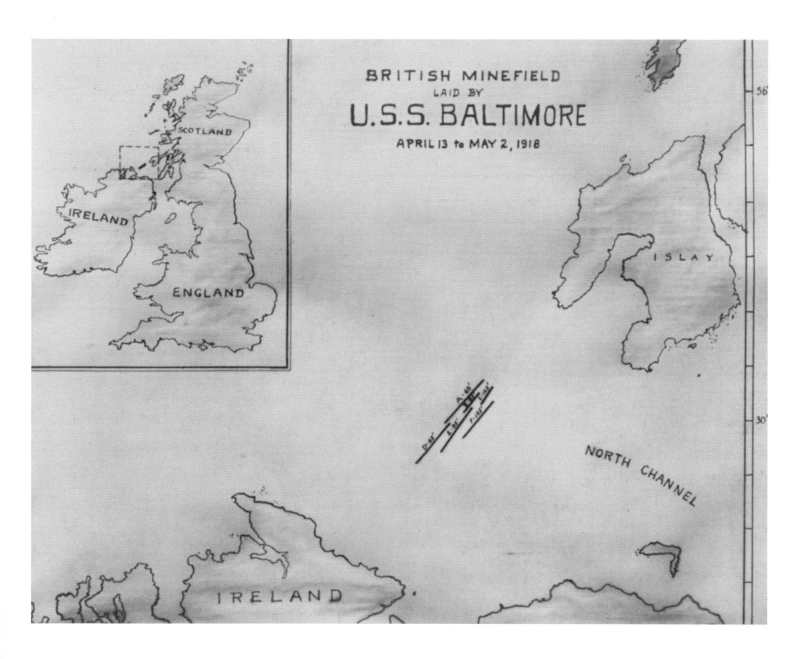

MINEFIELD LAID BY USS *BALTIMORE*, 1918
American entry into the First World War enhanced Allied naval strength, at a time when the Germans were becoming more powerful on land. American battleships strengthened the British Grand Fleet at Scapa Flow, American warships escorted Atlantic convoys and others laid mines to restrict submarine movements, as here between Scotland and Ireland. LEFT

THE STRATEGY OF PACIFIC CONFLICT, 19 FEBRUARY 1939, *SUNDAY NEWS* (NEW YORK) An overly-optimistic view of how Britain and the United States could limit Japanese naval action in the Pacific, by mounting a long-distance blockade, anchored at Singapore, Hawaii, and Dutch Harbor in the Aleutians. The map was originally printed on 3 October 1937 in response to Japanese expansionism in China. This map was updated and reprinted on 19 February 1939 in accordance with American plans for naval expansion. In practice, there is an acceptance of the vulnerability of the Philippines and Guam, while Wake Island is outside the Anglo-American blockade line, as is Midway. By the mid-1930s, more cautious American commanders had prevailed over the idea of a 'through ticket,' the plan for a rapid projection of American naval power to the Philippines. In the event, the British were unable to mount an effective blockade line, in part due to the strains of fighting Germany and Italy as well, and also as a result of the successful Japanese air attack on the *Prince of Wales* and *Republic* off eastern Malaya on 18 December 1941. In the naval battles of the Java Sea north of Java from 27 February to 1 March 1942, five cruisers were lost due to a lack of adequate Allied air power. **OPPOSITE**

The Second World War saw great demand for maritime information as naval operations took on an unprecedented combined range and intensity. Alongside the accumulation of information, it was necessary to disseminate what had been produced. Both reflected a situation of acute operational need. For example, the American landing force on the island of Guadalcanal in the south-western Pacific in 1942, the first sustained American counter-offensive, lacked adequate maps, including tidal maps, a problem that indicated the need for special amphibious landing maps. This was not the sole deficiency. The American naval raid on Wake Island in October 1943 faced the problem of inadequate charts for the surrounding waters. Such issues had to be surmounted and the resulting information disseminated. The scale of production of information was unprecedented: the US Naval Hydrographic Office printed more than forty million charts in the final year of the war. Other states could not match this scale, but also faced unprecedented demands.

The interwar years had suggested that the need for locational information would be more comprehensive and far more complex than had been anticipated prior to the First World War. In part, this reflected the anticipation that submarines would play a major role, as had not been appreciated in 1914. Another technology, however, posed a much more formidable challenge for planning and mapping. Naval air power had come to play a role in the First World War, notably in providing air cover against submarines. However, the use of aircraft carriers in the First World War had been very limited, while there had been no conflict between carriers. In contrast, by 1939, Britain, Japan, the USA and France had carriers, while the specifications of carrier aircraft had increased greatly in step with their land counterparts. There were still

Page 10

Long Distan

COLONIES, DEPENDENCIES and MANDATES

BRITISH JAPANESE
DUTCH PORTUGUESE
FRENCH UNITED STATES

NAVAL BASES AND HARBORS

AIR BASES

U.S. DEFENSES
PROPOSED ~ or to be ENLARGED
★ AIR BASES
● SUBMARINE BASES
▲ DESTROYER BASES
■ MINE BASES

BLOCKADE LINE CONTROLLED BY BRITISH FLEET

Longitude East from Greenwich

BLOCKADE

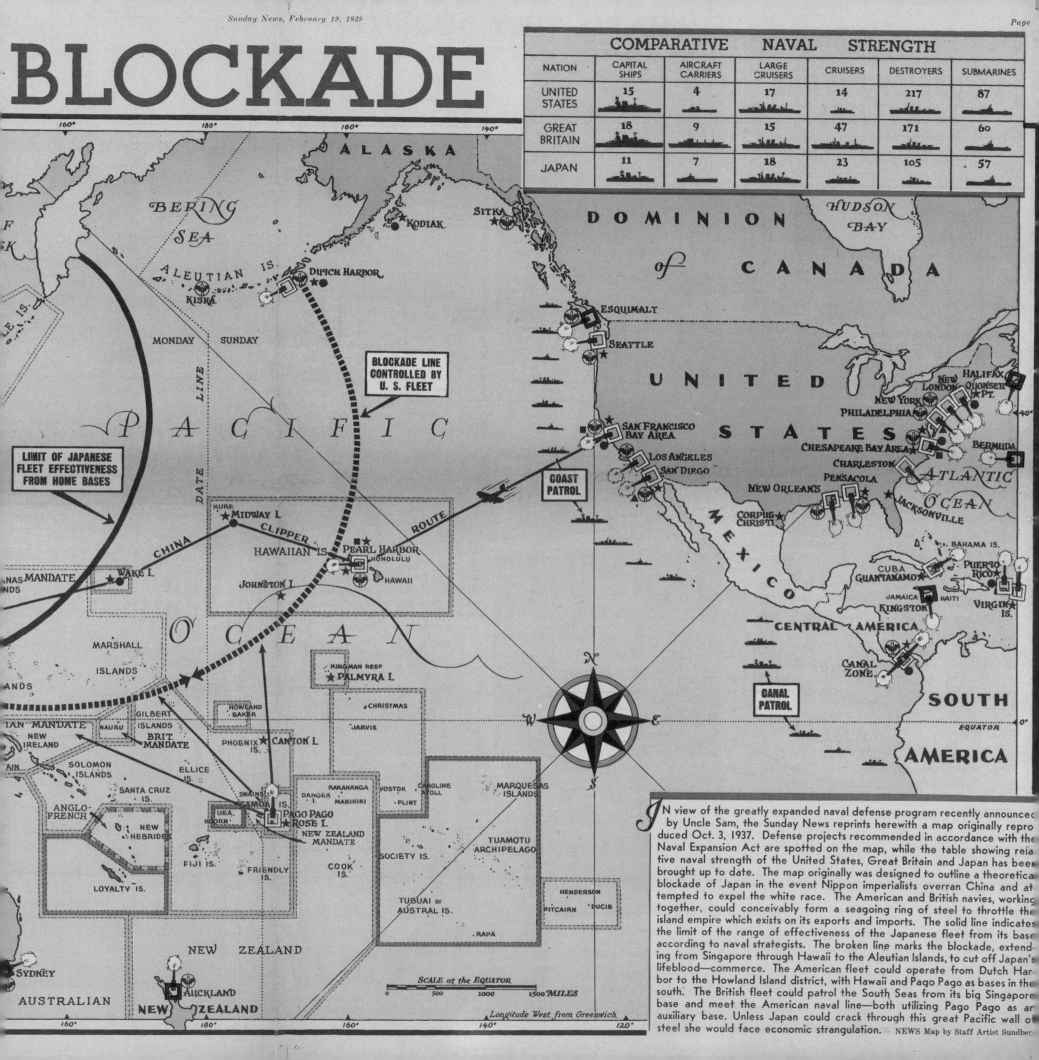

NATION	CAPITAL SHIPS	AIRCRAFT CARRIERS	LARGE CRUISERS	CRUISERS	DESTROYERS	SUBMARINES
UNITED STATES	15	4	17	14	217	87
GREAT BRITAIN	18	9	15	47	171	60
JAPAN	11	7	18	23	105	57

COMPARATIVE NAVAL STRENGTH

BLOCKADE LINE CONTROLLED BY U. S. FLEET

LIMIT OF JAPANESE FLEET EFFECTIVENESS FROM HOME BASES

COAST PATROL

CANAL PATROL

SCALE at the EQUATOR
0   500   1000   1500 MILES

Longitude West from Greenwich

IN view of the greatly expanded naval defense program recently announced by Uncle Sam, the Sunday News reprints herewith a map originally reproduced Oct. 3, 1937. Defense projects recommended in accordance with the Naval Expansion Act are spotted on the map, while the table showing relative naval strength of the United States, Great Britain and Japan has been brought up to date. The map originally was designed to outline a theoretical blockade of Japan in the event Nippon imperialists overran China and attempted to expel the white race. The American and British navies, working together, could conceivably form a seagoing ring of steel to throttle the island empire which exists on its exports and imports. The solid line indicates the limit of the range of effectiveness of the Japanese fleet from its base according to naval strategists. The broken line marks the blockade, extending from Singapore through Hawaii to the Aleutian Islands, to cut off Japan's lifeblood—commerce. The American fleet could operate from Dutch Harbor to the Howland Island district, with Hawaii and Pago Pago as bases in the south. The British fleet could patrol the South Seas from its big Singapore base and meet the American naval line—both utilizing Pago Pago as an auxiliary base. Unless Japan could crack through this great Pacific wall of steel she would face economic strangulation. NEWS Map by Staff Artist Sundber

RESHAPING THE ATLANTIC The scale and range of German submarine attacks threw the issues of convoy routes and escorts to the fore. The challenge and responses were different in the two world wars, not primarily due to technology, but as a result of the territory controlled by the two blocs. In 1940, the German occupation of France and Norway was crucial, but the British occupation of Iceland was a significant counter as it enhanced the potential for Allied escorts. A state under the Danish crown, Iceland declared independence when the Germans conquered Denmark in April 1940. To prevent a German intervention, the British landed troops there. The Atlantic was to be remapped when the US entered in the war in December 1941 and when Allied aircraft were granted access to the Azores in October 1943. OPPOSITE

major limitations for carrier aircraft, particularly operating at night or in cloud, but the prospect of carriers playing a major role, notably as part of combined-arms fleets, engaged much attention. It was assumed, as a result, that surface ships would have to protect themselves with anti-aircraft guns and aircraft cover.

UNPREDICTABILITY AND NAVAL PLANNING

The need to consider what war would be like if waged across the distances of the Pacific also engaged attention. The latter appeared a key prospect for the USA, as competition with Japan became more acute, and war with Japan dominated American naval planning from the 1920s. The Japanese had been aware from the 1900s that it might be a prospect. In addition, British naval planning, again from the 1920s, focused heavily on the possibility of war with Japan and the need, in consequence, to send a fleet to protect the British empire in Southeast Asia and Australasia, a fleet stationed in a new base at Singapore. Uncertainty and volatility were the key facets of a system in which planning and procurement increasingly focused on a war, the alignments and contours of which appeared highly unpredictable. The naval limitations diplomacy and treaties that followed the First World War, notably the Washington Naval Treaty of 1922, did not supersede rivalries and instead helped focus those between Japan and both America and Britain.

Alongside the stress on great-power naval competition that dominates the literature, there continued to be the use of naval power for policing. Much of this was imperial in character, as in the Persian Gulf where in 1910 an operation to find illicit arms in Dubai led to a raid by a British warship that caused fighting, while in 1921 the fort at Ajman and in 1925 that at Fujairah were bombarded by British

warships, in each case due to disputes over slavery. In 1927, British warships provided intimidation to back a settlement of a dispute between the rulers of Dubai and Sharjah, and in 1928 between Dubai and Iran. In 1932, the use of naval power helped gain landing rights for British aircraft. Such activity is too easily neglected. In terms of the information important for mapping, such activity helped ensure that new material on local maritime conditions was constantly being obtained.

Although some powers, especially Britain versus Germany, were in the same relationship throughout, from 1939 to 1945, this was not the case with most relationships. As a consequence, the tasking that the navies would have to confront was unclear, and thus the relative and contextual nature of strategy was highly dynamic. This was particularly the case for Britain as the German conquest of France in June 1940 deprived Britain of its major ally at the time and gave Germany French Atlantic bases, especially Brest, St Nazaire, La Rochelle and Bordeaux, from which Britain's trade routes across the Atlantic could be threatened by submarines, surface ships and aircraft. The surviving submarine pens serve as a powerful indication of the effort Germany made. Moreover, the earlier conquest of Norway provided Germany with further bases from which British trade could be threatened or attacked, and made it difficult to blockade Germany, as had been done in the First World War. The first-division naval powers at the time were Britain, the USA and Japan, with France, Germany and Italy making up the second-division powers. Britain suffered from Italy's entry into the war in June 1940 and, even more, from that of Japan in December 1941.

As a consequence, the Royal Navy, which had hoped to fight opponents separately and to benefit from allies, faced tremendous pressure. At the same time, the navy's strength helped deter the Germans from their planned

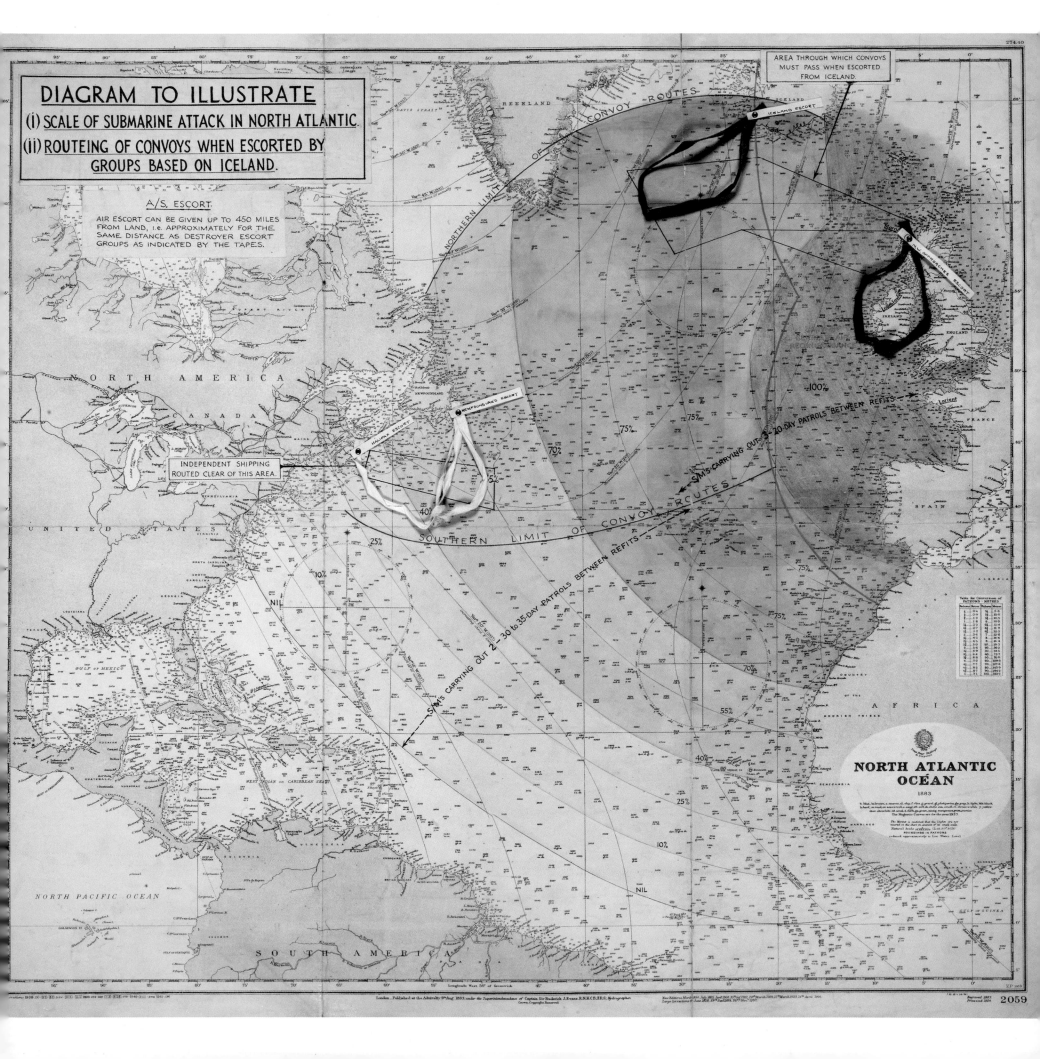

DIAGRAM TO ILLUSTRATE
(i) SCALE OF SUBMARINE ATTACK IN NORTH ATLANTIC.
(ii) ROUTEING OF CONVOYS WHEN ESCORTED BY
     GROUPS BASED ON ICELAND.

A/S. ESCORT.

AIR ESCORT CAN BE GIVEN UP TO 450 MILES
FROM LAND, i.e. APPROXIMATELY FOR THE
SAME DISTANCE AS DESTROYER ESCORT
GROUPS AS INDICATED BY THE TAPES.

AREA THROUGH WHICH CONVOYS
MUST PASS WHEN ESCORTED
FROM ICELAND.

INDEPENDENT SHIPPING
ROUTED CLEAR OF THIS AREA.

NORTH ATLANTIC
OCEAN
1883

2059

THE U-BOAT THREAT, 1940 British diagram of 1940 analysing U-boat threat to a convoy. Submarines were responsible for about 70 per cent of the Allied shipping destroyed by the Germans during the war, most of the rest falling victim to aircraft, mines and surface raiders. The British resorted to convoys much more rapidly than in the First World War. RIGHT

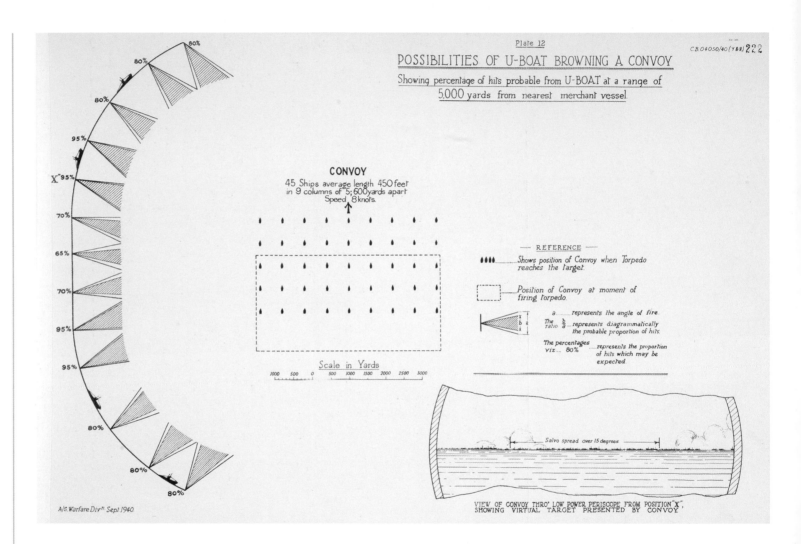

invasion of southern England in 1940. The struggle in the air was also very important, but even if the Germans had gained aerial superiority their navy would still have been vulnerable to British naval attack, especially at night.

CHALLENGES OF THE SECOND WORLD WAR
Pressure on the Royal Navy was accentuated by the extent to which her opponents made good use of newish technology, particularly submarines in the case of Germany and aircraft in that of Japan. As a result,

the British lost heavily in surface units, notably in 1941, and to a much larger extent than in the First World War. Equally, there were no battles comparable in scale or importance to Jutland, although there were many clashes with German surface warships, for example, the sinking of KMS *Bismarck* in 1941 after the German battleship had first sunk HMS *Hood*. Sent to attack Atlantic shipping, the *Bismarck* fell victim to British aircraft and surface bombardment, indicating a synergy between different forms of naval power that are often kept separate for purposes of analysis.

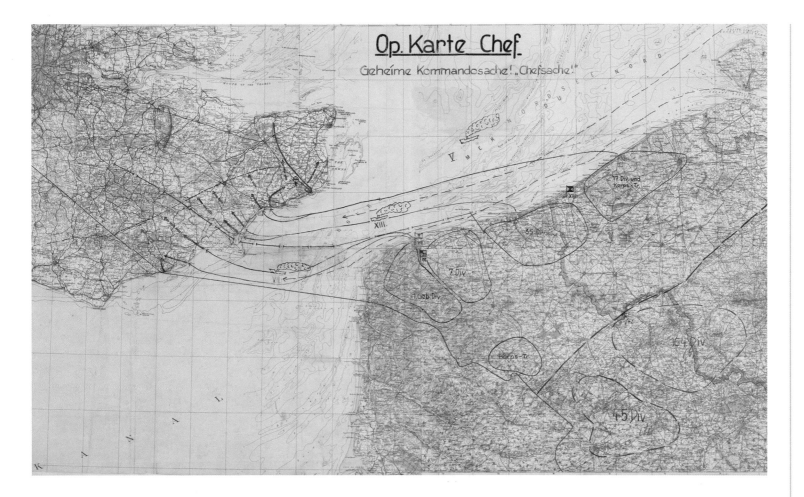

Op. Karte Chef
Geheime Kommandosache! "Chefsache!"

INVASION MAPPING, 1940 Planning map for Operation Sealion, the projected invasion of Britain. German preparations for an invasion of Britain included large-scale town plans marked with strategic locations, copies of Ordnance Survey maps, with overprints highlighting sites which were targets, the shoreline marked in terms of cliff, sand, steep and flat, and photographs to help with selected invasion beaches. In practice, aside from a lack of air superiority in 1940, the Germans lacked adequate experience or understanding of amphibious operations, as well as specialised landing craft. The towed Rhine barges they proposed to rely on could only manage a speed of three knots and would have failed to land a significant number of troops had any of them managed to reach England's south coast. The Royal Navy was probably strong enough to thwart invasion even had the *Luftwaffe* been more powerful, not least because of the limited night-time effectiveness of air power. The navy's vulnerability o German dive-bombers was an important factor, but these bombers were themselves vulnerable to British fighters and to antiaircraft guns. Moreover, the Germans lacked a torpedo-bomber capability. Even had the Germans landed troops, they probably would not have been able to sustain and reinforce them. **LEFT**

In the Mediterranean, there were battles between the British and Italian fleets, with the British victorious, notably in their carrier strike attack on Italian warships at Taranto in 1940 and in the Battle of Cape Matapan in March 1941. British success proved crucial to the dominance of British surface warships, and important to the course of hostilities in North Africa, only for that dominance to be sharply challenged by German aircraft and submarines. Success in individual battles or smaller engagements indicated strategic trends at sea but could not necessarily determine them. Despite Italian setbacks and defeats, including the Battles of Sirte in December 1941 and March 1942, the Italians were able to continue moving supplies to their forces in North Africa. In the face of serious losses to German air attack and submarines, the Royal Navy was able to continue operating in the Mediterranean, notably supplying the island of Malta, and managed to evacuate most of the British forces from mainland Greece and Crete in 1941. Moreover, Germany and Italy were not able to bring their naval power to bear on the British military position in Egypt in 1940–42. Indeed, in strategic terms, the Axis powers failed seriously, despite the availability of plentiful air power, submarines and nearby bases.

Month of SEPTEMBER
COASTAL COMMAND

SUBMARINE SIGHTINGS AND ATTACKS
AVERAGE DAY'S PATROLS
ANTI-SUBMARINE PATROLS
CONVOY ESCORTS
ANTI-INVASION PATROLS

3 Blenheims
0730-0858

1820/15
Hudson

4 Hudsons
1705-1940

1 Sunderland
1945-0100/12

1 Hudson
0545-0840

1927/16
Hudson

1 Spitfire
1030-1530

1230/16
Sunderland

EN 5

4 Hudsons
1600-1900

1515/5
Anson

WN 14

0920/29
Anson

OA 212

0745/9
Blenheim

2 Hudsons
0815-1115

OA 211

1730/13
Anson

2 Hudsons
0815-1100
1600-1845

HX 69A

0845/10
Hudson

1014
Hudson

1240/8
Hudson

1910/8
Hudson

0830/29
Hudson

1 Stranraer
2050-0140/12

MT

1 Sunderland
0800 - 1430

1727/9
Hudson

O.B. 211

2 Hudsons
0815-1145
1600-1930

SL 45

FS 76

FN 77

HX 69B

1312/2
Anson

1 Hudson
1000-1200

2 U/Boats.
1845/7
Hudson

OB 212

1 Hudson
2230/10 - 0230

1 Hudson
0215-0545
1945-2345

1 Hudson
2230-0300/12

FS 77

1 Spitfire
1630-1930

1 Spitfire
1460-1830

SC 2

1 Hudson
0400-0700
1915-2215

1 Blenheim
0915-1345

0520/2
Hudson

1 Hudson
0345-0650

OBM 212

1 Blenheim
0712-0918

FN 78

Blenheim
0452-0528
2020-2205

2 Ansons
0720-1015
1200-1435
1515-2015

1 Spitfire
0650-0840

1 Spitfire
0650-0900

1 Blenheim
2030-2230

CW 11

1 Spitfire
1700-2000

1 Sunderland
0745 - 1915

1 Blenheim
2030-2230

1 Anson
1930-2300

2 Blenheims
0650-0700

⊙ .... Submarines Sighted.
✛ .... „ Attacked.
———▶ .... A/S. Patrols.
——▶ .... Patrols *other than* A/S. and Escort.
∿∿∿ .... Convoy Escorts.

1 Spitfire
0850-1100

1217/5
Civil.

The pressure on the Royal Navy once Japan entered the war in December 1941 had a greater impact. Shore-based Japanese aircraft sank British ships off Malaya that month: HMS *Prince of Wales* became the first moving battleship sunk by aircraft. This defeat of a poorly commanded force ensured that the Royal Navy, which lacked the planned carrier support, would not be able to disrupt the Japanese conquest of Malaya and Singapore as it should have done. Subsequent Japanese successes in early 1942 in the Java Sea (over a British-Dutch-American-Australian fleet on 27 February–1 March), and off Sri Lanka (over the British in early April) had comparable effects in enabling the Japanese to invade Java and in ensuring that the British could not challenge Japan's conquest of Burma (Myanmar). These successes owed much to superior Japanese air power, which left Allied surface ships at a serious disadvantage. Seven Allied cruisers and one carrier were lost in these two operations.

PACIFIC ARENA

It was in the Pacific that naval warfare most frequently took the form of battle. From the Battle in the Coral Sea on 4–8 May 1942 onwards (a battle in which aircraft carriers engaged each other for the first time), competing carrier forces played a major role. At Coral Sea, the Japanese and Allied fleets were not in visual range, and the focus was on naval air power. Each side lost one carrier. As line of sight ceased to be significant, especially for range finding, the context for evaluating the situation changed and, with that, the need for plotting relative location and thus the material that could be mapped.

Being checked at Coral Sea led Japan to abandon the attempt to take Port Moresby in New Guinea by amphibious attack, instead focusing on an overland advance from northern New Guinea. The Japanese

now advanced eastwards in the Central Pacific. However, the USA won a key advantage at the Battle of Midway (4–7 June 1942), with the dive bomber attacks on 4 June leading to the sinking of four Japanese carriers and the loss of their maintenance crews as well as of pilots proving a major hit to the Japanese carrier skill base. Midway demonstrated that the problems of coordinating units and operations in one battle over a large body of water were formidable. This proved the case for both sides. At least, Midway Island provided the advantage of a fixed land point of reference. The Americans were successful, but the battle did not work out as either side had anticipated. A key aspect was that the Japanese battleships did not come into action, as the Japanese had sought – in large part because the American carriers took care to avoid coming too close.

The fate of individual operations should be put in the context of American industrial superiority. The American shipbuilding and aircraft manufacturing programmes ensured a major shift in the balance of resources against the Japanese, and this helped give the USA a key competitive advantage. As a result of their industrial capability, naval strength, and broad talent base, the Americans were able to play a major role in the Atlantic as well as taking responsibility for the Allied cause in the Pacific. There, American naval power supported offensives in the southwestern and Central Pacific, which put significant cumulative pressure on the Japanese.

At the same time, it was necessary to develop techniques and skills in order to benefit from this advantage. For example, in the Pacific, resupply provided a key organisational capability, one that enabled the Americans to keep their fleets at sea. In addition, radar-directed fire proved important in battleship clashes, notably at night off Guadalcanal

COASTAL COMMAND MONTHLY RECORD, NOVEMBER 1940 The challenge from German submarines became much worse after the German conquests of Norway and France. Coastal Command's record of its activities noted changes in the range of operations. The German submarines operating north of Ireland were aiming for Britain's transoceanic trade en route to Glasgow and Liverpool. The RAF played a key role in resisting the assault. Most U-boat 'kills' of shipping were made by attack on the surface, which rendered Allied sonar less effective. In contrast, aircraft forced U-boats to submerge where their speed was slower, their reliance on battery-powered electric motors affected cruising range and the number of days that could be spent at sea, and it was harder to maintain visual contact with targets. **OPPOSITE**

in late 1942, giving the Americans an important advantage while also robbing the Japanese of their former dominance of battle capability at night. The American ineptitude at night fighting early in the Pacific War was due largely to the apparent effectiveness of radar in the pre-war fleet exercises which gave senior American naval commanders a false understanding of the limitations of the technology, especially when employed by hastily trained personnel. These clashes illustrated the range of warship types involved in the conflict.

Aside from surface shipping, submarines came to play an important role. They enabled the Americans, operating from Pearl Harbor and from Freemantle in West Australia, to dislocate the Japanese war economy in what was the most successful submarine campaign in history and one that is largely underrated outside the USA. This campaign was made despite the long-time inadequacy of American torpedoes and involved significant losses. British and Dutch submarines also played a major role in this campaign.

In contrast, the Germans failed to do the same against the Allies in the Atlantic. In large part, this was a result of the Allied success in developing successful anti-submarine techniques, but Allied shipbuilding was also very important. The comparative dimension is instructive. Japan had entered the war with a relatively small merchant marine, especially tankers. Despite major efforts, Japan did not build sufficient merchantmen to provide the strategic depth (in the shape of capacity to take losses) that the Allies had in the face of German attacks. Japan also failed to create a serious convoy escort force. American submarines often operated in areas where the Allied forces were not yet in control of the sky, so Japanese escort destroyers would have been useful.

AMPHIBIOUS ATTACKS

While naval warfare in these forms dominated attention, it also proved crucial in the shape of supporting amphibious attacks. These were of only limited significance for Germany and, even more, the Soviet Union. For Japan they were of limited importance in its war against China after 1937, although valuable in southern China. However, the Japanese had specialised landing craft and their conquest of the Dutch East Indies and the western Pacific in 1941–42 totally depended on amphibious operations.

So did American war making, both in the Pacific and in Europe, and its British counterpart in Europe, and even more so from late 1942. The capability of this war making was altered by specialised powered landing craft which made assaults on beaches more practical, thus reducing the need to capture ports, notably in the D-Day attacks on Normandy in June 1944. The Allies landed on beaches and did not need to capture Cherbourg or Le Havre in order to operate. The creation and use of a prefabricated harbour, which was transported across the Channel, helped the Allies to establish and hold the Normandy beachhead.

Allied skill was cumulative, with invasions of Morocco and Algeria (1942), Sicily (1943), mainland Italy (1943), northern and southern France (1944) each showing the benefit of experience, although lessons were not always learned, as the problems the Allies encountered with Italian landings in 1943 and 1944 indicated. Amphibious operations required the collection of data, such as hydrographic and beach conditions, which could be exploited by naval and amphibious units. The work by the British COPPs (Combined Operations Assault Pilotage Parties) was crucial to the success of the D-Day attacks and the amphibious assaults in the Mediterranean.

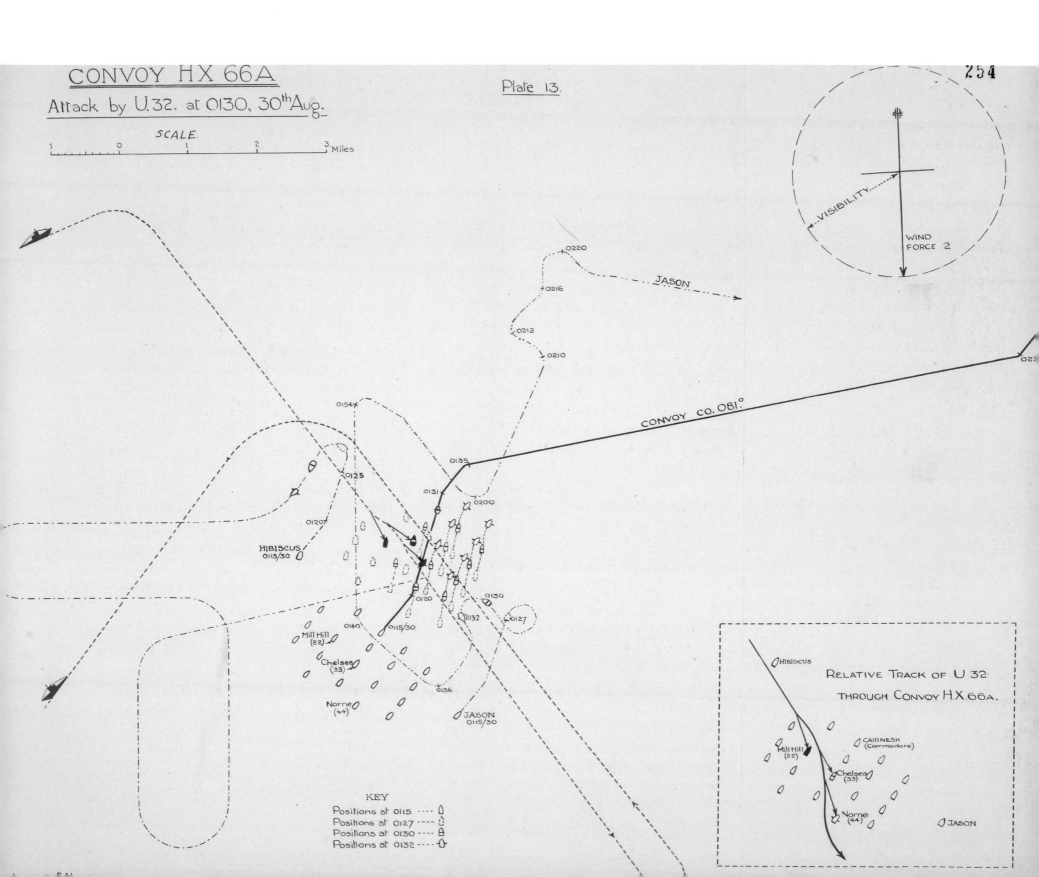

CONVOY HX 66A

Attack by U.32. at 0130, 30ᵗʰAug.

Plate 13.

254

SCALE.

1 0 1 2 3 Miles

VISIBILITY

WIND FORCE 2

0220

0216 JASON

0212

0210

CONVOY CO. 081°

0154

0135

0125

0131

0200

0120

HIBISCUS
0115/30

0130

0132

0127

0140

0115/30

Mill Hill
(22)

Chelsea
(33)

0136

Norne
(44)

JASON
0115/30

KEY

Positions at 0115 ·····
Positions at 0127 ----
Positions at 0130 ----
Positions at 0132 ----

HIBISCUS

RELATIVE TRACK OF U 32
THROUGH CONVOY H.X. 66A.

CAIRNESK
(Commodore)

Mill Hill
(22)

Chelsea
(33)

Norne
(44)

JASON

MAR PICCOLO

TRENTO   BOLZANO

DESTROYERS

PTA RONDINELLA

PORTO MERCANTILE

CASTEL ST ANGELO

MAR

GRANDE

CRUISERS

NETS

NETS

DUILIO

CAVOUR

LITTORIO

LITTORIO

DUILIO

SCANNO

ISOLOTTO
SAN
PIETRO

IL POSTO

ISOLOTTO
SAN PAOLO

DIGA DI TARANTOLA

CAVOUR

OIL PIPE-LINE
JETTY

FLOATING
DOCK

DIGA DI SAN VITO

N

TORPEDO-DROPPING POSITIONS
ANTI-AIRCRAFT BATTERIES
BARRAGE BALLOONS

CAPO SAN VITO

OIL STORAGE
DEPOT

0   ½   1   ½   2   3   4   5   LED·VERNO

ETRES

Naval supporting fire was important in the suppression of coastal defences and in helping resist German counterattacks, as at Salerno in 1943 and Anzio in 1944, while naval cover was crucial in preventing disruption at sea from enemy action. The Japanese repeatedly sought to use fleet action to destroy or damage American landing forces and were thwarted, increasingly with heavy casualties, especially in the Battle of Leyte Gulf in 1944 and the Battle of Okinawa in 1945. The Germans and Italians did not make comparable efforts, but, had they done so, they would have been defeated.

The course of the war demonstrated the advantages of gaining naval superiority, and the extent to which its use then proved an important war winner. By late 1940, the threat posed by the Italian navy had been largely overcome, in part with the movement northwards of Italian warships after the successful British carrier-based torpedo-bomber attack at the base of Taranto. This attack prefigured that, on a greater scale, made by the Japanese at Pearl Harbor in December 1941.

By the end of the summer of 1942, the Japanese navy had suffered a crushing defeat at Midway, which greatly lessened its offensive capability, while the balance of effectiveness in the Atlantic submarine war was increasingly moving towards Britain and America. These verdicts became even clearer over the following year, with the Americans eventually successful off Guadalcanal over the Japanese and German submarines largely defeated in early 1943 – the key campaign in the Battle of the Atlantic. This ensured that the Allies would be able to build up their military strength in Britain preparatory to an invasion of France. An earlier invasion would have been dangerous given the submarine challenge. The sinking by the Vichy French of their fleet at Toulon in November 1942 prevented the Germans from seizing the ships.

Moreover, this fleet was not deployed to oppose the Anglo-American invasion of Northwest Africa in Operation Torch. This was one of the most impressive naval operations of all times, given the size of the fleet, the distance to travel, the untested nature of much of the American fleet and a lack of British and American experience in co-operating. There was not the same resistance on land as with the Japanese in the Pacific, but the logistics and the overall organisation were very impressive for the time, as was the considerable daring involved.

Allied successes in 1943 looked towards the scale, pace, range and impact of Allied amphibious operations in 1944. In 1943, Italy abandoned Germany, which altered the ratio of naval advantage in the Mediterranean even further in the Allied favour.

AIDING THE SOVIETS

The Axis powers were unable to counter the extent to which the Allies were able to use the oceans to move resources and, increasingly, to shape the course of the war. This was crucially the case in terms of keeping Britain in the war in the face of German submarine attacks, and putting pressure on the Germans in Western Europe. This pressure affected the dispositions of the German army, reducing the number of troops deployed to face the Soviet army. Thus, the Allied invasion of Sicily in 1943 helped lead the Germans to abandon their Kursk offensive in Russia. More generally, supplies sent to the Soviet Union by sea, notably from the USA to Siberia and from Britain to northern Russia, increased the mobility of Soviet units, especially with the provision of American jeeps and trucks. In December 1941, one sixth of the Soviet heavy tanks in the Battle of Moscow had been supplied and shipped by the British. They could be provided as the Germans had

NAVAL AIR ATTACK The British seized the initiative on 11 November, 1940, by means of a successful surprise night attack on the Italian naval base of Taranto by twenty-one torpedo bombers launched from the carrier *Illustrious* stationed 180 miles away. Two aircraft were lost, but three battleships and a cruiser were torpedoed with the British adopting the technique of shallow-running the torpedoes. In terms of the damage inflicted per attacking aircraft, this raid was more effective than the Japanese raid on Pearl Harbor the following year which it encouraged. In response to the raid, the Italians withdrew units from Taranto northward and thus lessened the vulnerability of British maritime routes and naval forces in the Mediterranean, notably by increasing the problems of concentrating Italian naval forces and maintaining secrecy. For Britain, Taranto served as a crucial boost to public and government morale at a time of international isolation and serious German air attacks in the Blitz. OPPOSITE

WEEKLY DIAGRAM of U-BOAT W

SEPT. 30TH – OCT. 27TH 1940.

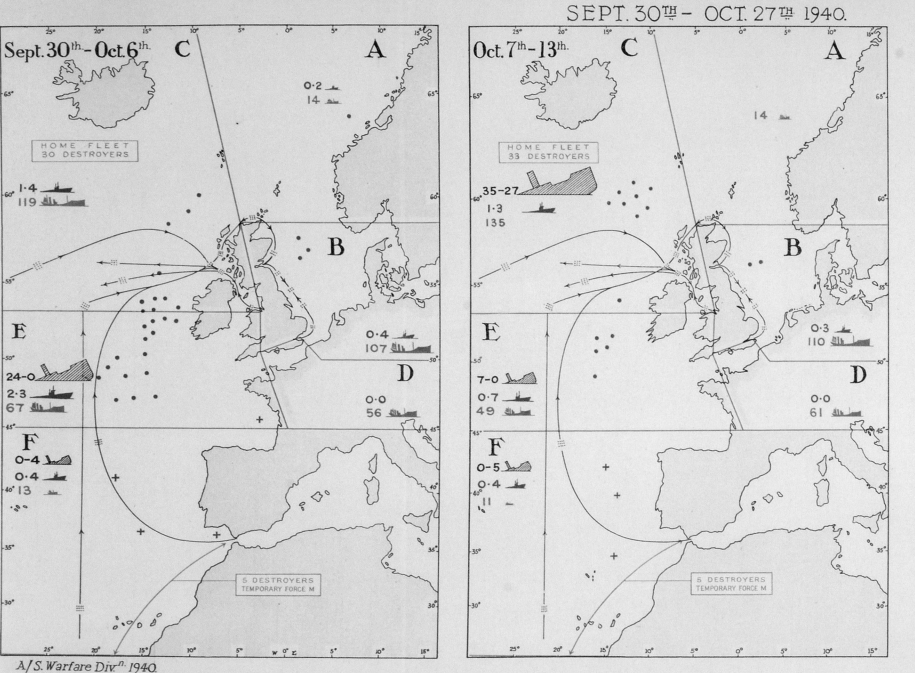

A/S. Warfare Div.ᵑ 1940.

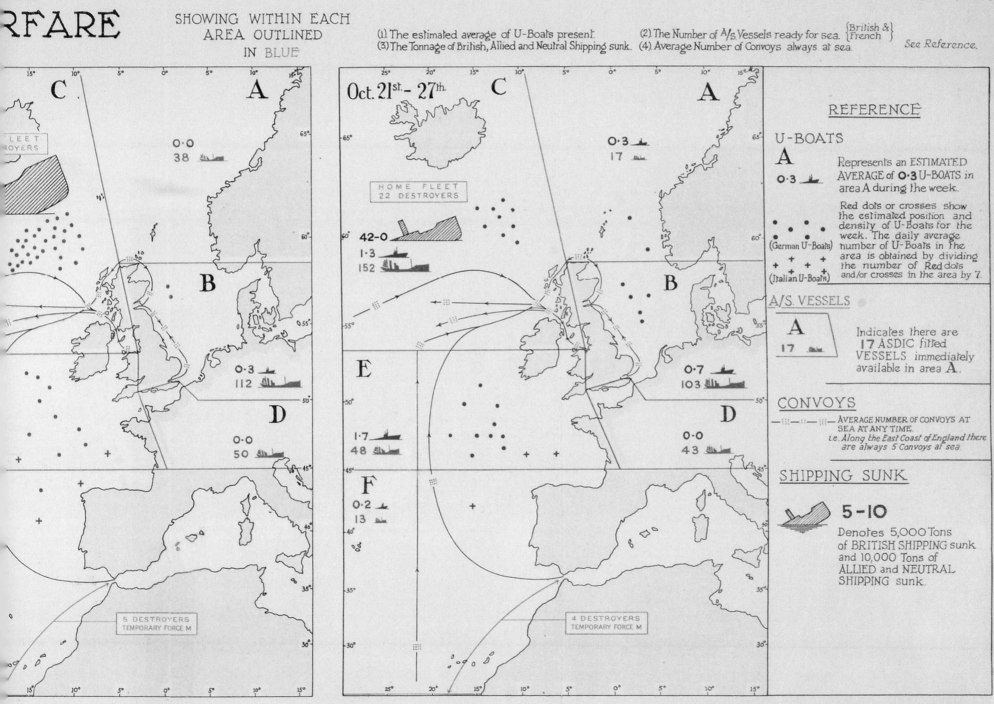

RFARE

SHOWING WITHIN EACH
AREA OUTLINED
IN BLUE

(1) The estimated average of U-Boats present.
(3) The Tonnage of British, Allied and Neutral Shipping sunk.

(2) The Number of A/s Vessels ready for sea. {British & French}
(4) Average Number of Convoys always at sea.

See Reference.

259

C.B. 04050/40 (9110).

Oct. 21st - 27th

REFERENCE

U-BOATS

A
0·3 Represents an ESTIMATED AVERAGE of 0·3 U-BOATS in area A during the week.

(German U-Boats)
(Italian U-Boats)

Red dots or crosses show the estimated position and density of U-Boats for the week. The daily average number of U-Boats in the area is obtained by dividing the number of Red dots and/or crosses in the area by 7.

A/S VESSELS

A
17 Indicates there are 17 ASDIC fitted VESSELS immediately available in area A.

CONVOYS

———— AVERAGE NUMBER OF CONVOYS AT SEA AT ANY TIME.
i.e. Along the East Coast of England there are always 5 Convoys at sea.

SHIPPING SUNK

5-10 Denotes 5,000 Tons of BRITISH SHIPPING sunk and 10,000 Tons of ALLIED and NEUTRAL SHIPPING sunk.

HOME FLEET
22 DESTROYERS

5 DESTROYERS
TEMPORARY FORCE M

4 DESTROYERS
TEMPORARY FORCE M

abandoned their plans to invade Britain. American
trucks were crucial to the Soviet westward advances
in 1943–45.

The major, and costly, effort made by the Royal
Navy to assist the Soviet Union was an aspect of the
failure of the Soviet navy to fulfil Stalin's pre-war
hopes and plans. The naval command structure had
been badly hit by political intervention, most
conspicuously with the bloody purges of the late
1930s, which had also hit naval shipbuilding and
support. In 1941, the German attack hit the Soviet
fleet hard in the Baltic Sea and, both there and in the
Black Sea, the Soviets were forced to withdraw in the
face of German land advances and seizure of ports.
The Soviet fleet was still able to pay a significant role,
especially in withdrawing forces from besieged Odessa
in 1941 and in supplying Sevastopol when under
eventually successful German siege in 1942.

Soviet submarines, some provided by Britain,
attacked German convoys in Norwegian waters which
brought supplies to the German land forces that
unsuccessfully sought to capture Murmansk. Soviet
submarines also attacked Swedish ships transporting
iron ore to Germany across the Baltic Sea. However,
on the whole, the Soviet fleet failed to play a
significant operational role other than in support of
land forces, as with operations on the Kerch peninsula.
Concern about losses led Stalin to urge caution. In
1945, when the Soviet Union attacked Japan, the
Soviets successfully used amphibious operations to
attack the Japanese in Sakhalin and the Kurile Islands.

In supplying the Soviet Union, the Royal Navy
faced not only harsh Arctic seas, but also repeated
German attacks, notably by submarines and aircraft.
German surface warships were also concentrated in
Norwegian waters. The battlecruiser KMS *Scharnhorst*,
tracked by radar, was sunk by shells and torpedoes in

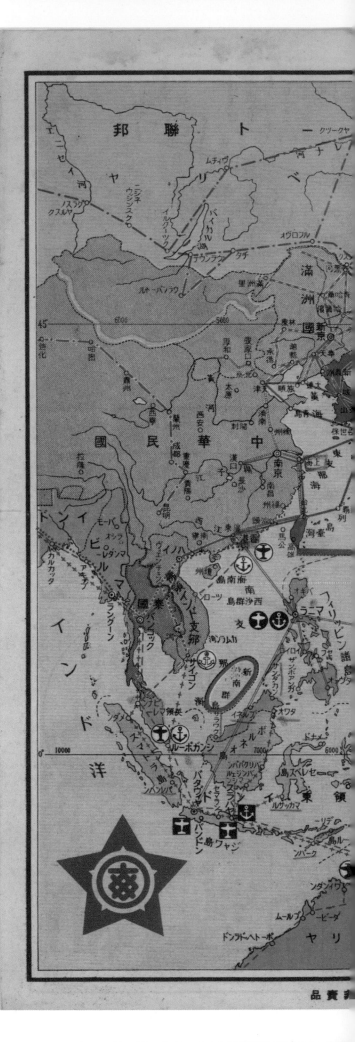

護れ太平洋要圖　大阪毎日新聞

精版印刷所 大阪市西區江戸堀上ニ四ノ三二　須古清 大阪市西宮松下町九五
編輯兼發行人　發行所 大阪市北區堂島上二丁目三六 大阪毎日新聞社　昭和十六年八月二十日發行　不許複製

the Battle of North Cape on 26 December 1943, while the battleship KMS *Tirpitz* was sunk by British bombers near Tromsö on 12 November 1944.

NAVAL CONTRIBUTION

To a degree, the naval element in Axis defeat has been overlooked due to the argument that the Soviet army defeated Germany and the atom bombs did the same for Japan. Indeed, the American and British navies largely lost the battles of reputation as the reasons for success in the world war were subsequently (and at the time) contested. For example, the air force gained the credit for stopping Germany from invading Britain in 1940, rather than the naval strength that was the key element. Similarly, the success of the American submarine campaign against Japan was underplayed.

There was also tension within navies, tension in which recent history was linked to disagreements over doctrine and procurement, and to the pursuit of

careers. This was notably so with an emphasis on carriers at the expense of surface firepower, especially from battleships. In this, the postwar situation was read back into the assessment of the war.

In practice, the defeat of Germany, Italy and Japan owed much to the Allies achieving mastery of the oceans. This mastery provided a vital strategic advantage, as so often with naval capability. The advantage was both narrowly military, notably in terms of moving naval units and in supporting amphibious operations, and also more broadly so. In the latter case, the ability to articulate war economies and to support a widespread, in this case global, alliance system was dependent on maritime routes and capacity, and thus on the naval ability to protect them and to challenge those of opponents. The continued supply of food and raw materials was crucial in keeping Britain in the war. The supply of munitions and food to the Soviet Union was also significant.

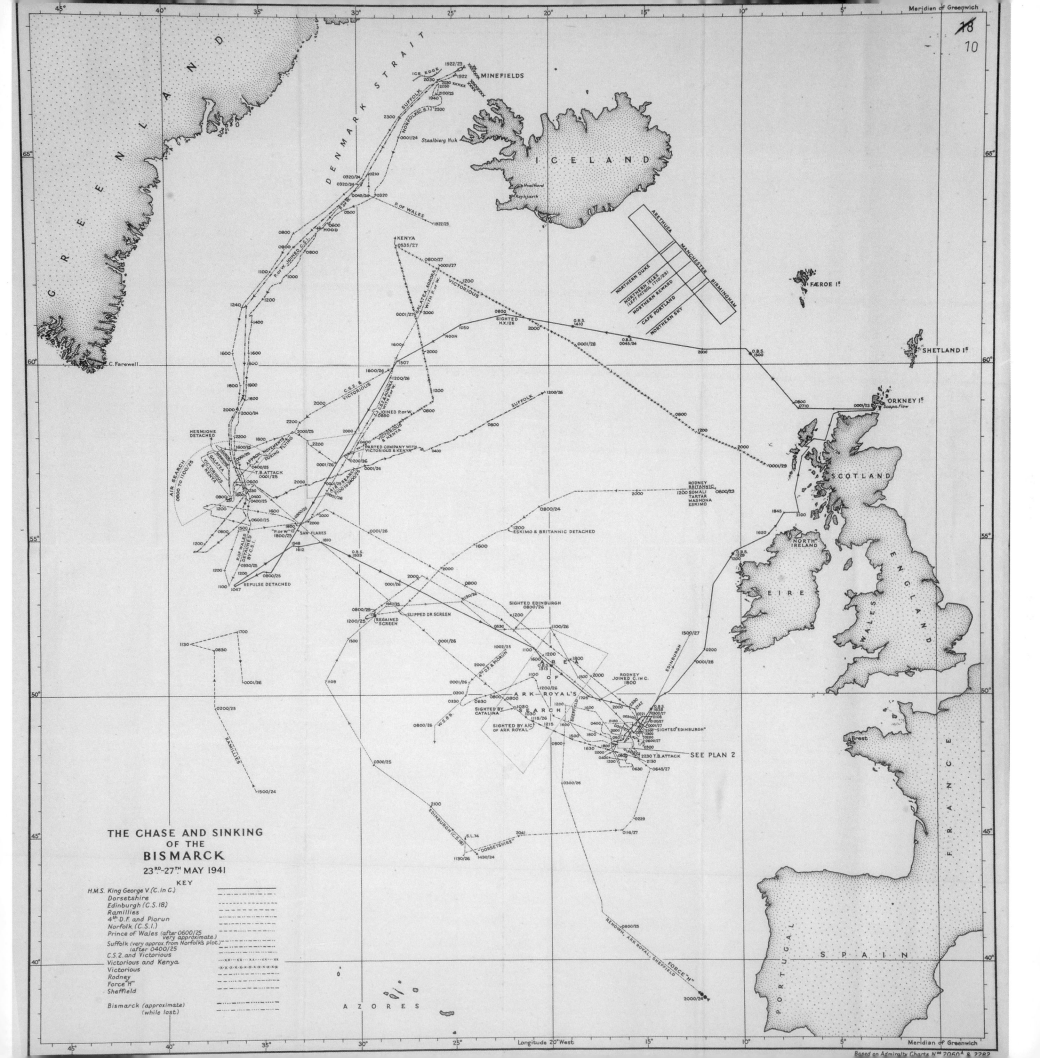

THE CHASE AND SINKING
OF THE
BISMARCK
23RD-27TH MAY 1941

KEY

H.M.S. King George V. (C. in C.)
Dorsetshire
Edinburgh (C.S.18)
Ramillies
4TH D.F. and Piorun
Norfolk (C.S.I.)
Prince of Wales (after 0600/25 very approximate)
Suffolk (very approx from Norfolk's plot.) (after 0400/25)
C.S.2. and Victorious
Victorious and Kenya
Victorious
Rodney
Force "H"
Sheffield

Bismarck (approximate) (while lost)

JAPANESE MAP OF PEARL HARBOR, 1941 This Japanese map showing Pearl Harbor was found in a captured midget submarine launched to attack ships during the attack on the base. The Japanese used five two-man, twenty-four metre (seventy-eight and a half feet) 46 Ko-Gata submarines in the attack. They were carried to the area by I-type submarines. Neither class of submarine achieved anything during the operation, which hit the reputation of the Japanese submarine force as a whole. The Japanese midget submarines were more successful when used at the expense of British warships at Diego Suarez, Madagascar, in May 1942.

RIGHT

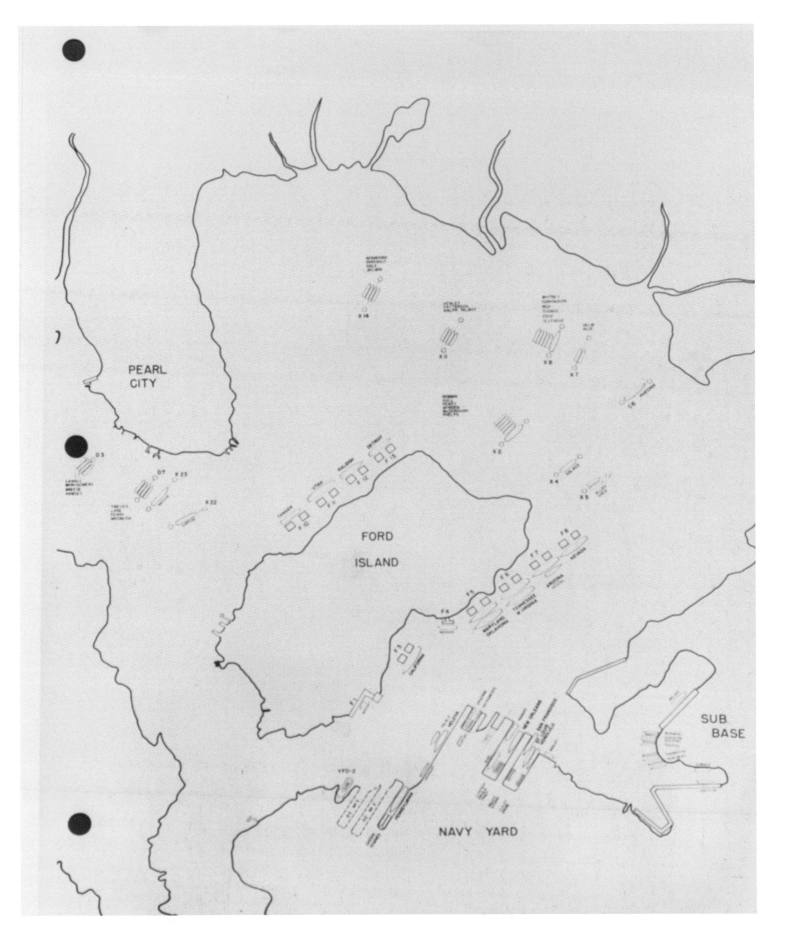

PEARL
CITY

FORD
ISLAND

NAVY YARD

SUB.
BASE

THE ATTACK ON PEARL HARBOR, 7 DECEMBER 1941
US map showing the location of
American warships when the Japanese
attacked. 353 aircraft from six Japanese
carriers completely destroyed two
American battleships and damaged five
more, while, in an attack on the naval air
station at Kaneohe Bay, nearly 300
American aircraft were destroyed or
damaged on the ground. This was an
operational-tactical success, but a
strategic and grand strategic failure.
The attack revealed grave deficiencies
in Japanese (and American) planning, as
well as in the Japanese war machine. The
Japanese suffered from late preparation
and a lack of practice. The Japanese
target-prioritisation scheme was poor,
attack routes conflicted, and the torpedo
attack lacked simultaneity. The damage
to America's battleships (some of which
were to be salvaged and used anew)
forced an important shift in American
naval planning toward an emphasis on the
carriers, which, despite Japanese
expectations, were not in Pearl Harbor
when it was attacked. The devastating
nature of this surprise attack encouraged
a rallying around the American
government. LEFT

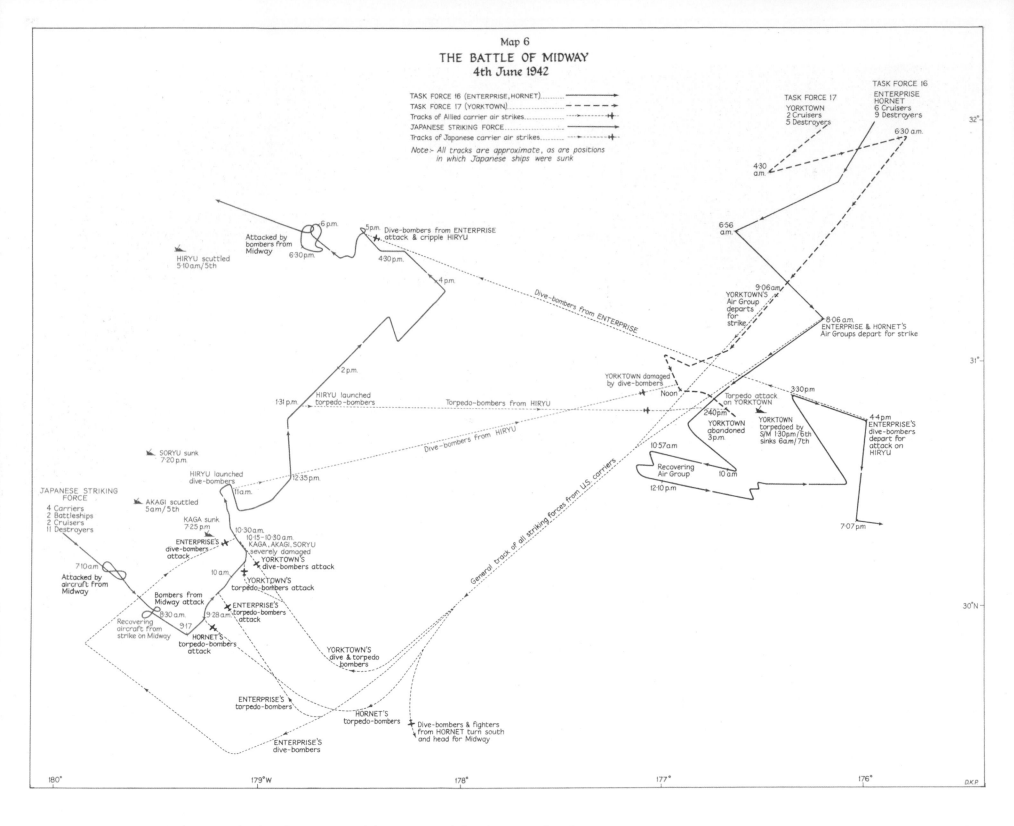

Map 6
THE BATTLE OF MIDWAY
4th June 1942

TASK FORCE 16 (ENTERPRISE, HORNET)..........
TASK FORCE 17 (YORKTOWN)..........
Tracks of Allied carrier air strikes..........
JAPANESE STRIKING FORCE..........
Tracks of Japanese carrier air strikes..........

*Note:- All tracks are approximate, as are positions
in which Japanese ships were sunk*

TASK FORCE 17
YORKTOWN
2 Cruisers
5 Destroyers

TASK FORCE 16
ENTERPRISE
HORNET
6 Cruisers
9 Destroyers

6.30 a.m.

4.30 a.m.

6.56 a.m.

Attacked by bombers from Midway
6 p.m.
5 p.m. Dive-bombers from ENTERPRISE attack & cripple HIRYU
HIRYU scuttled 5.10 a.m./5th
6.30 p.m.
4.30 p.m.

9.06 a.m. YORKTOWN'S Air Group departs for strike

Dive-bombers from ENTERPRISE
4 p.m.

8.06 a.m. ENTERPRISE & HORNET'S Air Groups depart for strike

2 p.m.

YORKTOWN damaged by dive-bombers
Noon

3.30 p.m.

HIRYU launched torpedo-bombers
1.31 p.m.
Torpedo-bombers from HIRYU

Torpedo attack on YORKTOWN
2.40 p.m.
YORKTOWN abandoned
YORKTOWN torpedoed by S/M 1.30pm/6th sinks 6a.m./7th

4.4 p.m. ENTERPRISE'S dive-bombers depart for attack on HIRYU

Dive-bombers from HIRYU

SORYU sunk 7.20 p.m.

HIRYU launched dive-bombers
12.35 p.m.
11 a.m.

10.57 a.m.
Recovering Air Group
10 a.m.
12.10 p.m.

7.07 pm

General track of all striking forces from U.S. carriers

JAPANESE STRIKING FORCE
4 Carriers
2 Battleships
2 Cruisers
11 Destroyers

AKAGI scuttled 5 a.m./5th
KAGA sunk 7.25 p.m.

10.30 a.m.
10.15–10.30 a.m. KAGA, AKAGI, SORYU severely damaged
ENTERPRISE'S dive-bombers attack
YORKTOWN'S dive-bombers attack

7.10 a.m.
Attacked by aircraft from Midway

10 a.m.
YORKTOWN'S torpedo-bombers attack

Bombers from Midway attack
8.30 a.m.
9.28 a.m.
ENTERPRISE'S torpedo-bombers attack

Recovering aircraft from strike on Midway
9.17
HORNET'S torpedo-bombers attack

YORKTOWN'S dive & torpedo bombers

ENTERPRISE'S torpedo-bombers

HORNET'S torpedo-bombers

Dive-bombers & fighters from HORNET turn south and head for Midway

ENTERPRISE'S dive-bombers

180°   179°W   178°   177°   176°

32°
31°
30°N

D.K.P.

THE BATTLE OF MIDWAY, 4 JUNE 1942 Midway was a naval-air battle of unprecedented scale, that reflected the effectiveness of the combination of fighter support with carriers (in defence) and of fighters and bombers (in attack). The Americans encountered serious problems in the battle, and contingency and chance played a major role in the fighting. The Japanese lost all four of their heavy carriers present, as well as many aircraft. There was no opportunity for the Japanese to use their battleships, as the American carriers prudently retired before their approach. Midway demonstrated the power of carriers, but also their serious vulnerability, not least if, like the Japanese, they had poor damage-control practices.   ABOVE

JAPANESE PROPAGANDA MAP OF 1942 This map provides a highly misleading account of the impact and range of Japanese power. Japanese aircraft are shown ranging widely in the Pacific, including off Sydney. Moreover, the Japanese impact in the Central Pacific was more limited than is shown. ABOVE

The Pacific—Vital Area for United States and Japan

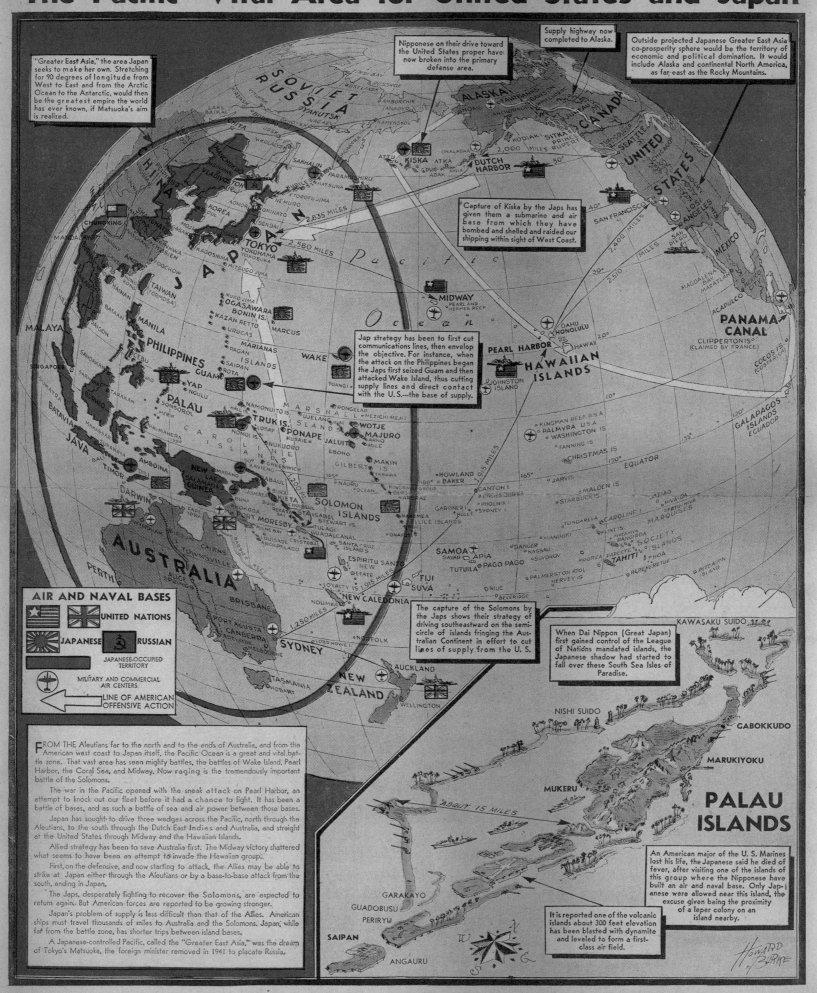

"Greater East Asia," the area Japan seeks to make her own. Stretching for 90 degrees of longitude from West to East and from the Arctic Ocean to the Antarctic, would then be the greatest empire the world has ever known, if Matsuoka's aim is realized.

Nipponese on their drive toward the United States proper have now broken into the primary defense area.

Supply highway now completed to Alaska.

Outside projected Japanese Greater East Asia co-prosperity sphere would be the territory of economic and political domination. It would include Alaska and continental North America, as far east as the Rocky Mountains.

Capture of Kiska by the Japs has given them a submarine and air base from which they have bombed and shelled and raided our shipping within sight of West Coast.

Jap strategy has been to first cut communications lines, then envelop the objective. For instance, when the attack on the Philippines began the Japs first seized Guam and then attacked Wake Island, thus cutting supply lines and direct contact with the U. S.—the base of supply.

The capture of the Solomons by the Japs shows their strategy of driving southeastward on the semi-circle of islands fringing the Australian Continent in effort to cut lines of supply from the U. S.

When Dai Nippon (Great Japan) first gained control of the League of Nations mandated islands, the Japanese shadow had started to fall over these South Sea Isles of Paradise.

AIR AND NAVAL BASES

UNITED NATIONS
JAPANESE
RUSSIAN
JAPANESE-OCCUPIED TERRITORY
MILITARY AND COMMERCIAL AIR CENTERS
LINE OF AMERICAN OFFENSIVE ACTION

FROM THE Aleutians far to the north and to the ends of Australia, and from the American west coast to Japan itself, the Pacific Ocean is a great and vital battle zone. That vast area has seen mighty battles, the battles of Wake Island, Pearl Harbor, the Coral Sea, and Midway. Now raging is the tremendously important battle of the Solomons.

The war in the Pacific opened with the sneak attack on Pearl Harbor, an attempt to knock out our fleet before it had a chance to fight. It has been a battle of bases, and as such a battle of sea and air power between those bases.

Japan has sought to drive three wedges across the Pacific, north through the Aleutians, to the south through the Dutch East Indies and Australia, and straight at the United States through Midway and the Hawaiian Islands.

Allied strategy has been to save Australia first. The Midway victory shattered what seems to have been an attempt to invade the Hawaiian group.

First, on the defensive, and now starting to attack, the Allies may be able to strike at Japan either through the Aleutians or by a base-to-base attack from the south, ending in Japan.

The Japs, desperately fighting to recover the Solomons, are expected to return again. But American forces are reported to be growing stronger.

Japan's problem of supply is less difficult than that of the Allies. American ships must travel thousands of miles to Australia and the Solomons. Japan, while far from the battle zone, has shorter trips between island bases.

A Japanese-controlled Pacific, called the "Greater East Asia," was the dream of Tokyo's Matsuoka, the foreign minister removed in 1941 to placate Russia.

PALAU ISLANDS

An American major of the U. S. Marines lost his life, the Japanese said he died of fever, after visiting one of the islands of this group where the Nipponese have built an air and naval base. Only Japanese were allowed near this island, the excuse given being the proximity of a leper colony on an island nearby.

It is reported one of the volcanic islands about 300 feet elevation has been blasted with dynamite and leveled to form a first-class air field.

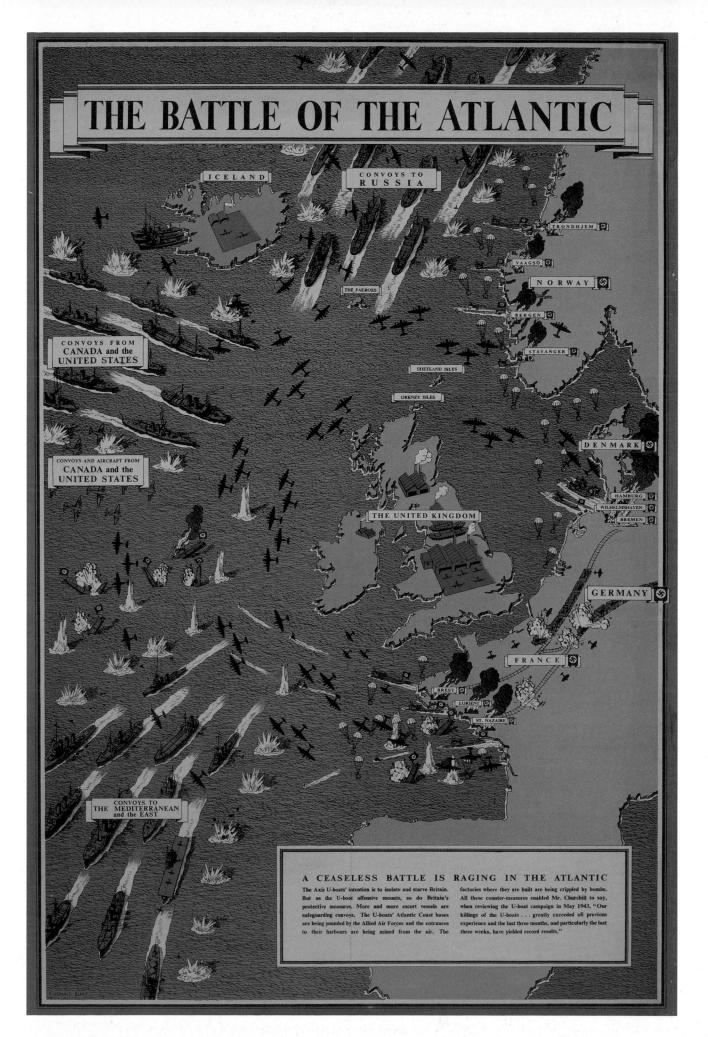

THE BATTLE OF THE ATLANTIC

ICELAND

CONVOYS TO
RUSSIA

TRONDHJEM

VAAGSÖ

NORWAY

THE FAEROES

BERGEN

STAVANGER

CONVOYS FROM
CANADA and the
UNITED STATES

SHETLAND ISLES

ORKNEY ISLES

CONVOYS AND AIRCRAFT FROM
CANADA and the
UNITED STATES

DENMARK

HAMBURG
WILHELMSHAVEN
BREMEN

THE UNITED KINGDOM

GERMANY

FRANCE

BREST
LORIENT
ST. NAZAIRE

CONVOYS TO
THE MEDITERRANEAN
and the EAST

A CEASELESS BATTLE IS RAGING IN THE ATLANTIC

The Axis U-boats' intention is to isolate and starve Britain. But as the U-boat offensive mounts, so do Britain's protective measures. More and more escort vessels are safeguarding convoys. The U-boats' Atlantic Coast bases are being pounded by the Allied Air Forces and the entrances to their harbours are being mined from the air. The factories where they are built are being crippled by bombs. All these counter-measures enabled Mr. Churchill to say, when reviewing the U-boat campaign in May 1943, "Our killings of the U-boats . . . greatly exceeded all previous experience and the last three months, and particularly the last three weeks, have yielded record results."

F. DONALD BLAKE

EDUCATING THE PUBLIC, NOVEMBER 1942 A colour map published in the *Los Angeles Examiner* in order to provide a general geopolitics of the Pacific War, to explain naval strategy and to locate the struggle for the Palau Islands. In July 1942, the Australian War Cabinet cabled Churchill: 'Superior seapower and airpower are vital to wrest the initiative from Japan and are essential to assure the defensive position in the southwest Pacific Area.' American success in building up naval strength combined with the development and use of important organisational advantages, notably in logistics. At the time of this map the Americans were not in a position to mount the offensive action indicated by the white arrows. **OPPOSITE**

THE BATTLE OF THE ATLANTIC, 1943, BY FREDERICK DONALD BLAKE A vivid account, but a misleading one. Far from being crowded in this fashion, there was a marked discrepancy between the roughness of the weather in the vast Atlantic, the small size of the ships and submarines and the problems of visibility. Changing goals were also significant. For example, U-boat warfare in the Atlantic was reduced in late 1941 as submarines were moved to Norwegian and Mediterranean waters in order to attack Allied convoys to the Soviet Union and to deny the Mediterranean to Allied shipping. **LEFT**

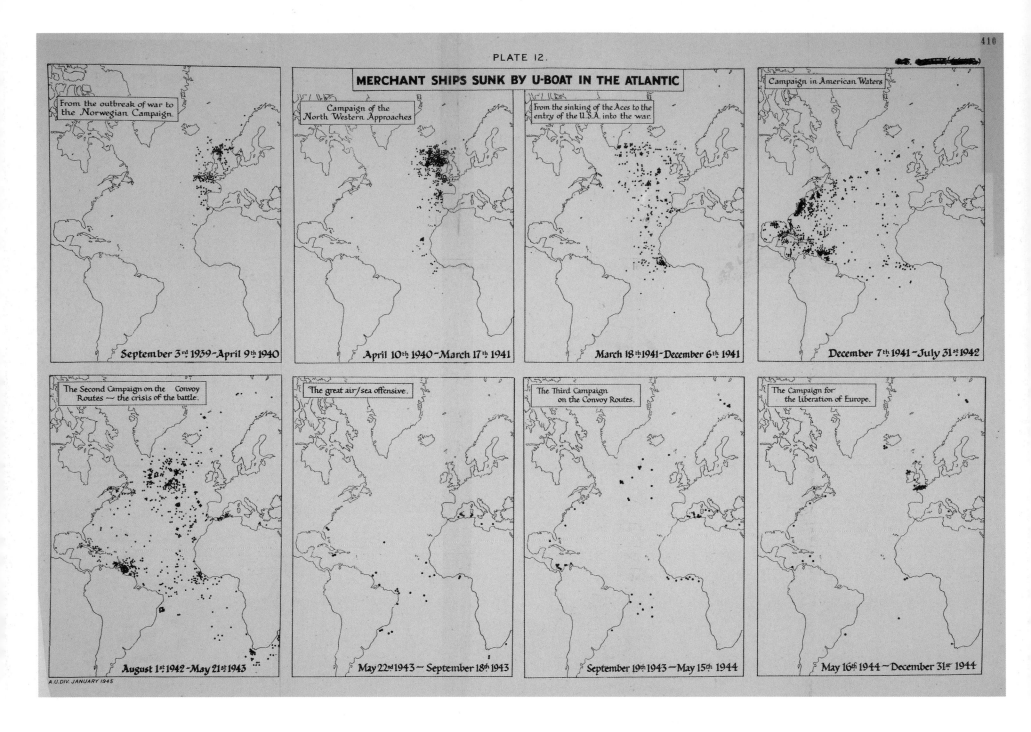

PLATE 12.

MERCHANT SHIPS SUNK BY U-BOAT IN THE ATLANTIC

From the outbreak of war to the *Norwegian Campaign*.

September 3rd 1939 – April 9th 1940

Campaign of the North Western Approaches

April 10th 1940 – March 17th 1941

From the sinking of the Aces to the entry of the U.S.A. into the war.

March 18th 1941 – December 6th 1941

Campaign in American Waters

December 7th 1941 – July 31st 1942

The Second Campaign on the Convoy Routes — the crisis of the battle.

August 1st 1942 – May 21st 1943

The great air/sea offensive.

May 22nd 1943 – September 18th 1943

The Third Campaign on the Convoy Routes.

September 19th 1943 – May 15th 1944

The Campaign for the Liberation of Europe.

May 16th 1944 – December 31st 1944

A.U.DIV. JANUARY 1945

THE BATTLE OF THE ATLANTIC, 1939–44 A British chronological breakdown of the stages of the Battle of the Atlantic, showing losses of Allied merchant ships. Allied success was crucial to the provision of imports to feed and fuel Britain, as well as to the build-up of military resources there that was necessary for D-Day. In the end the U-boats had only an operational capability for their strategic threat was thwarted. Mass was a key element. For example, the production of escort destroyers began in June 1943, and the USA built over 300 of them. Escort carriers were introduced and then many were built. In the last stage of the war, the U-boats, concentrated in Norway, focused on inshore British waters. Construction maintained the overall numbers of U-boats, obliging the Allies to continue to devote considerable naval resources to escort duty and antisubmarine warfare. Given Allied, especially American shipbuilding capacity, this did not prevent other uses of Allied naval power. Bombing delayed the construction of a new, faster class of submarine, which did not become operational until April 1945. The Allies were also developing new tactics and technology. **ABOVE**

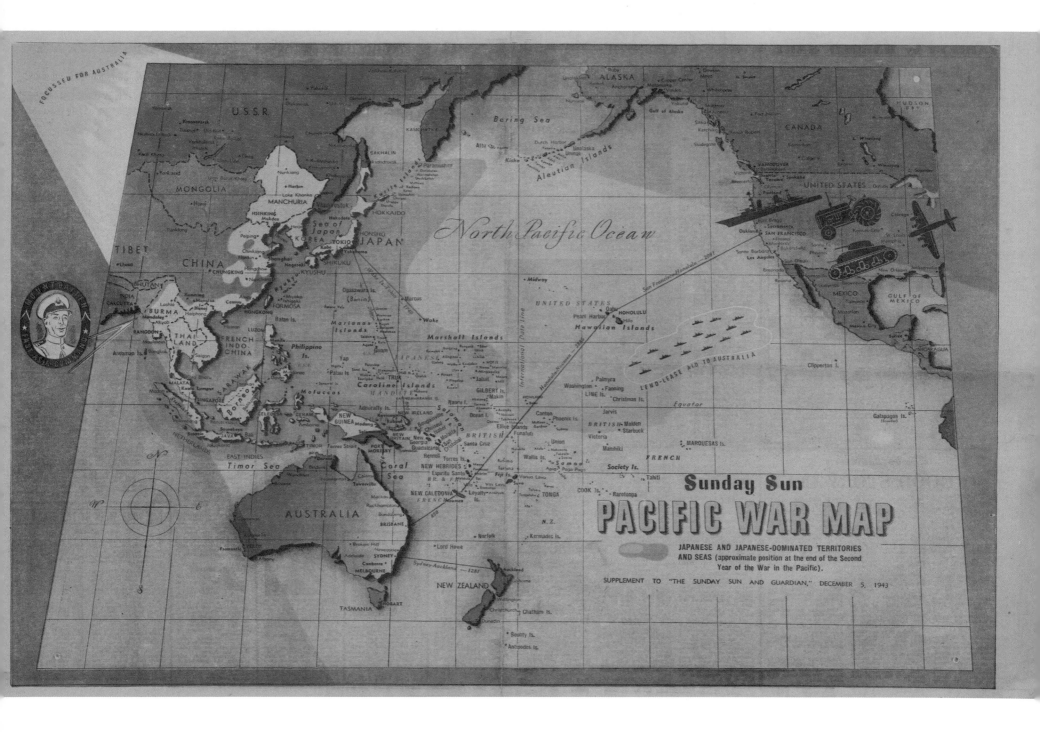

FOCUSSED FOR AUSTRALIA

Sunday Sun
PACIFIC WAR MAP

JAPANESE AND JAPANESE-DOMINATED TERRITORIES
AND SEAS (approximate position at the end of the Second
Year of the War in the Pacific).

SUPPLEMENT TO "THE SUNDAY SUN AND GUARDIAN," DECEMBER 5, 1943

MAP OF THE PACIFIC DISTRIBUTED AS A SUPPLEMENT TO SYDNEY'S *SUNDAY SUN AND GUARDIAN*, **5 DECEMBER 1943** This newspaper map shows Japanese and Japanese-dominated seas. The significance of the south-west Pacific to the maritime route from San Francisco to Brisbane emerges clearly. The United States appears as an arsenal of democracy. In the previous month, the Americans had opened up a new axis of advance in the central Pacific, capturing key atolls in the Gilbert Islands in hard-fought amphibious attacks on the islands of Makin and Taravoa. This success helped prepare the way for operations against the Marshall Islands in early 1944. In the south-west Pacific a force of American cruisers and destroyers, covering the landing on the island of Bougainville in the Solomons on 1 November 1943, beat off an attack that night by a smaller Japanese squadron, with losses to the latter in the first battle fought entirely by radar. **ABOVE**

THE WAR ON THE U-BOATS. THE SINKING OF

GERMAN AND ITALIAN SUBMARINES A range of
elements played a role in the defeat of
the U-boats, notably improved resources,
tactics, and strategy. More powerful
depth charges, effective ahead-throwing
guns, better sonar detection equipment,
increased use of shipborne radar, the
application of signals intelligence, and
the use of air power all played a role.
Alongside such incremental measures,
there was also the experience of
synergies due to operating together. In
early 1944, the Germans fitted snorkel
devices to their U-boats, which allowed
them to charge their batteries while
submerged and to start and run their
engines underwater. However, the snorkel
did not permit running the diesels at
sufficient depth to avoid detection from
the air. The improvement in Allied escort
capability outweighed U-boat advances,
and the Allied tonnage sunk per U-boat
fell. RIGHT

PLATE NO.10

GERMAN & ITALIAN

• Kills by Surface Craft (including Submarines)

From the outbreak of war to
the Norwegian Campaign.

September 3rd. 1939 ~ April 9 th. 1940

The Campaign of the
North Western Approaches.

April 10th. 194

The Second Campaign on the
Convoy Routes ~ the crisis of the battle.

August 1st. 1942 ~ May 21st. 1943

The great air/sea offensive.

May 22nd

A.U.D. February 1945

OATS DESTROYED (A & B ASSESSMENTS)

C.B.04050/45(1)

Kills~ by Aircraft   • Kills by combined attack of Surface Craft & Aircraft.

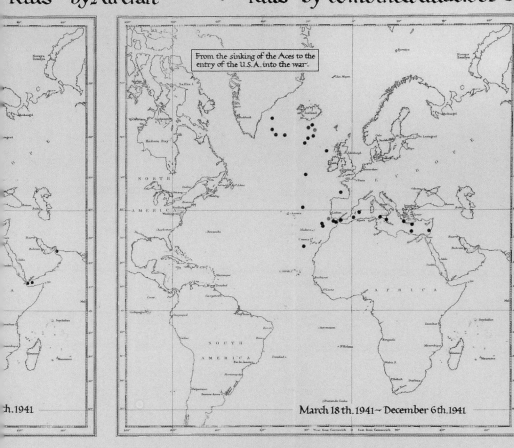

From the sinking of the Aces to the entry of the U.S.A. into the war.

March 18th. 1941 ~ December 6th. 1941

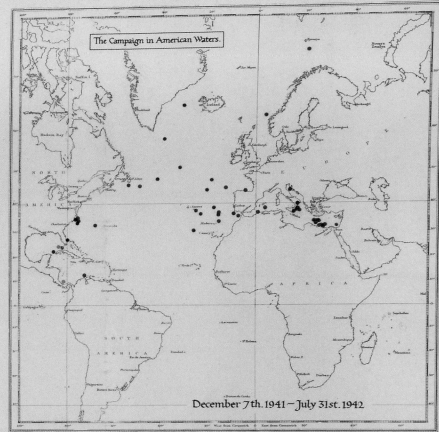

The Campaign in American Waters.

December 7th. 1941 ~ July 31st. 1942

ber 18th. 1943

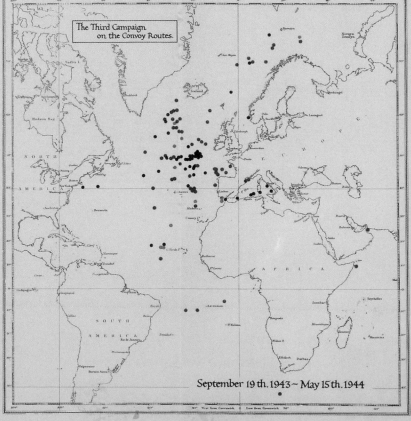

The Third Campaign on the Convoy Routes.

September 19th. 1943 ~ May 15th. 1944

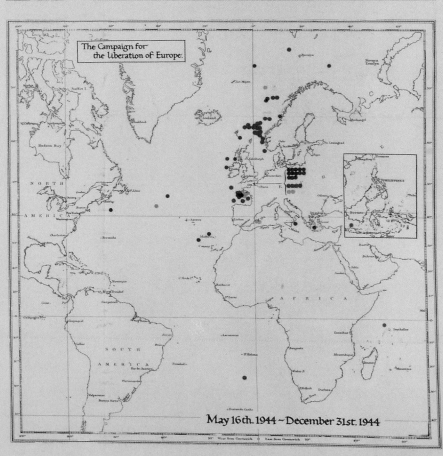

The Campaign for the liberation of Europe.

May 16th. 1944 ~ December 31st. 1944

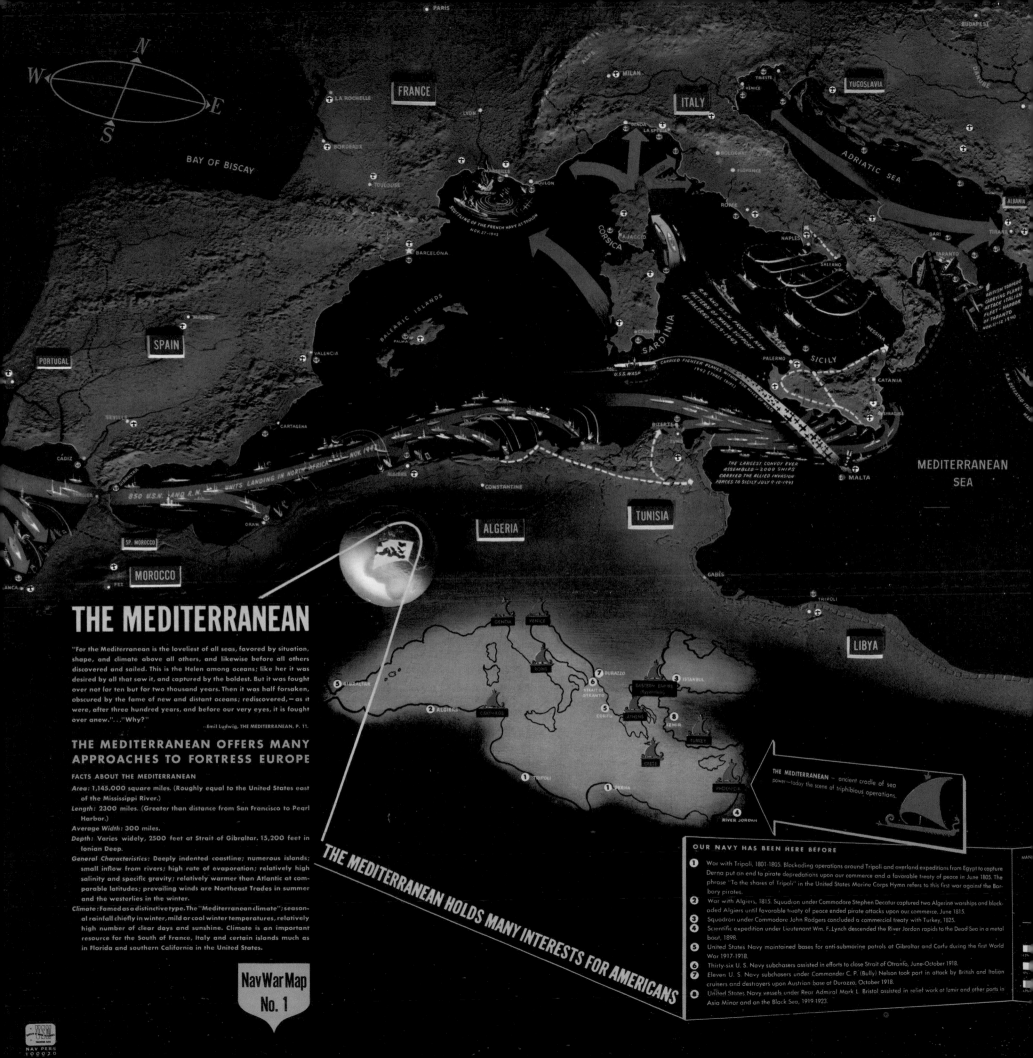

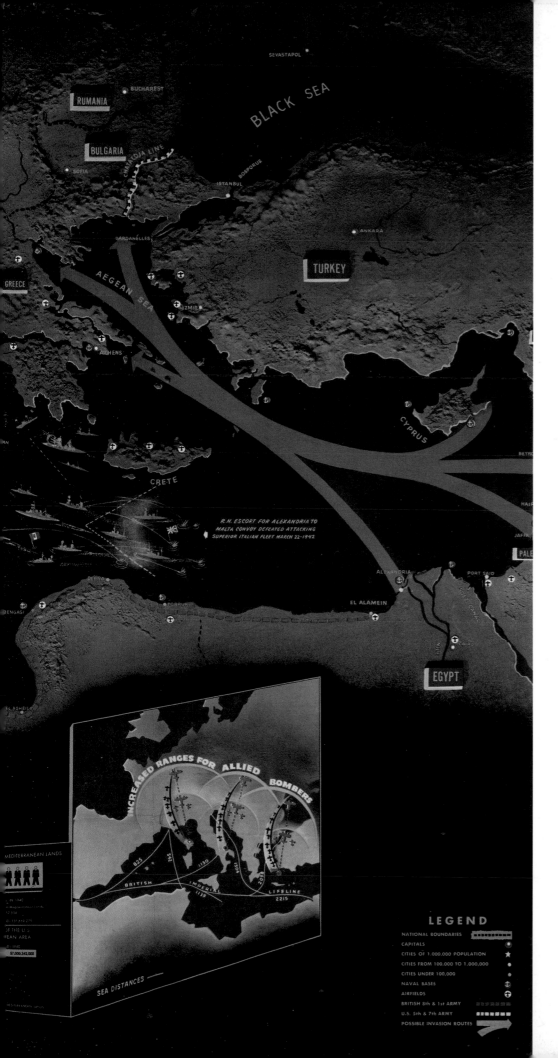

THE MEDITERRANEAN, 1944 The Educational Services Division of the Bureau of Naval Personnel of the American navy produced a series of maps informing staff and public about significant wartime events in particular areas, with an emphasis on the American role. This ensures that the map highlights Operation Torch and the subsequent invasions of Italy. Torch, the invasion of Morocco and Algeria in November 1942, had revealed a lack of appropriate equipment, doctrine and experience. Many of the faults had been rectified by the Anglo-American invasion of Sicily in July 1943 (which involved 3,600 ships), not least through proper reconnaissance for,

and organisation of, the process of landing. This was to be the second largest amphibious operation of the war in Europe, D-Day in Normandy being the first. The landings were more sophisticated than Operation Torch. There were appropriate ship to shore techniques, notably amphibious pontoon causeways and trained beach parties. Anglo-American naval bombardment and carrier support were important to the landings at Salerno in September 1943, Anzio in February 1944, and southern France in August 1944. In the last, 887 warships, including nine carriers, five battleships and twenty-one cruisers supported the 1,370 landing craft. LEFT

HAND DRAWN MAP BY AN AMERICAN MARINE SHOWING LOCATIONS OF FORCES LANDING ON SAIPAN There is a handwritten list of ships involved in the battle and their classification numbers on the verso. Landing on 15 June, 1944, the American Marines took heavy casualties. The determination of the Japanese resistance ensured that nearly the entire garrison of 27,000 men died resisting attack. The Japanese mounted a strong defence in the jungle-covered mountain terrain, but the island fell on 9 July, and the Japanese government resigned nine days later.

RIGHT

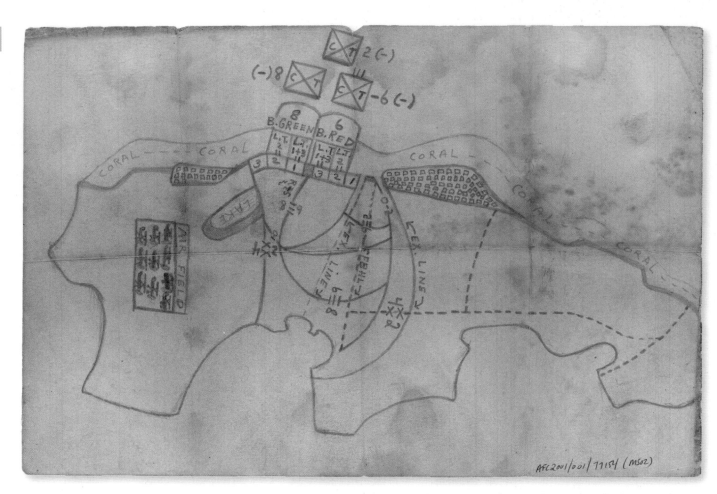

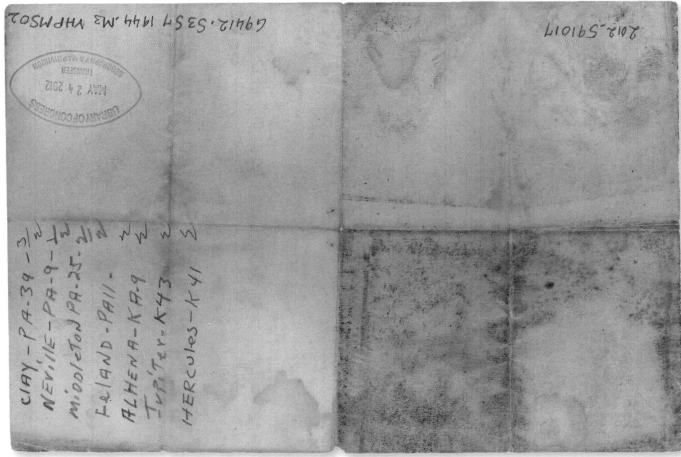

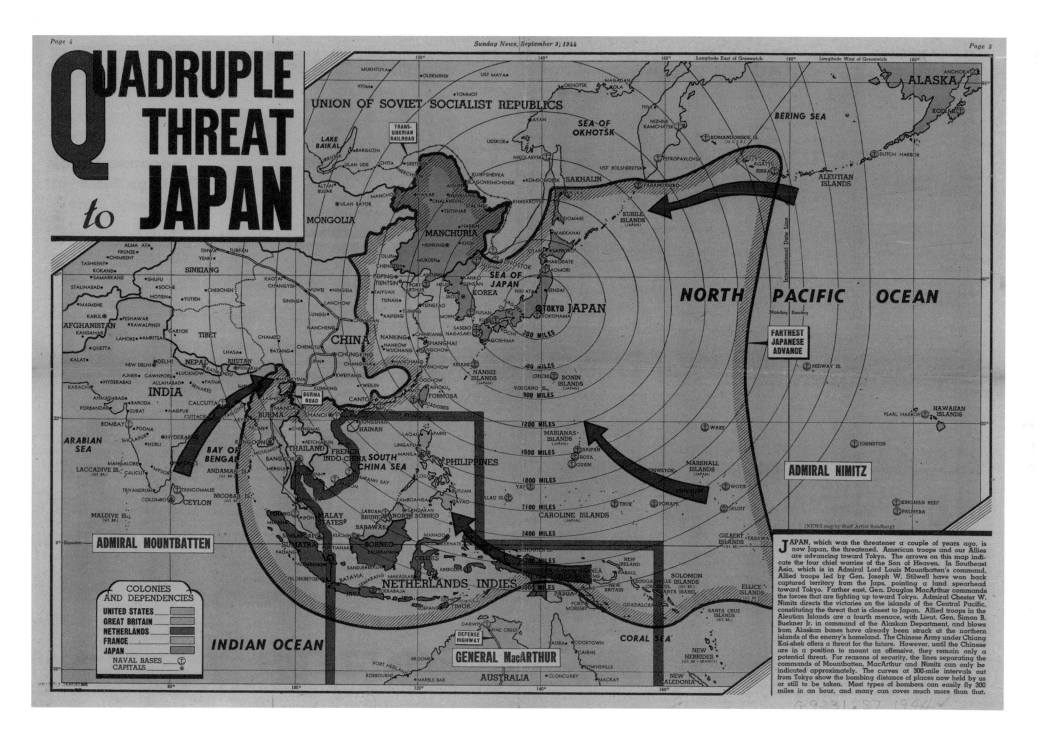

QUADRUPLE THREAT to JAPAN

UNION OF SOVIET SOCIALIST REPUBLICS

NORTH PACIFIC OCEAN

ADMIRAL NIMITZ

ADMIRAL MOUNTBATTEN

GENERAL MacARTHUR

INDIAN OCEAN

COLONIES AND DEPENDENCIES
UNITED STATES
GREAT BRITAIN
NETHERLANDS
FRANCE
JAPAN
NAVAL BASES
CAPITALS

JAPAN, which was the threatener a couple of years ago, is now Japan, the threatened. American troops and our Allies are advancing toward Tokyo. The arrows on this map indicate the four chief worries of the Son of Heaven. In Southeast Asia, which is in Admiral Lord Louis Mountbatten's command, Allied troops led by Gen. Joseph W. Stilwell have won back captured territory from the Japs, pointing a land spearhead toward Tokyo. Farther east, Gen. Douglas MacArthur commands the forces that are fighting up toward Tokyo. Admiral Chester W. Nimitz directs the victories on the islands of the Central Pacific, constituting the threat that is closest to Japan. Allied troops in the Aleutian Islands are a fourth menace, with Lieut. Gen. Simon B. Buckner Jr. in command of the Alaskan Department, and blows from Alaskan bases have already been struck at the northern islands of the enemy's homeland. The Chinese Army under Chiang Kai-shek offers a threat for the future. However, until the Chinese are in a position to mount an offensive, they remain only a potential threat. For reasons of security, the lines separating the commands of Mountbatten, MacArthur and Nimitz can only be indicated approximately. The curves at 300-mile intervals out from Tokyo show the bombing distance of places now held by us or still to be taken. Most types of bombers can easily fly 300 miles in an hour, and many can cover much more than that.

QUADRUPLE THREAT TO JAPAN, SUNDAY NEWS (NEW YORK), 3 SEPTEMBER, 1944 The map shows the Allies converging on Japan, with radial distances centred on Tokyo in order to indicate bombing distances. In practice, there was no threat of an advance from the Kuriles or of a 'land spearhead toward Tokyo' from Burma via China. Instead, the Japanese made major gains from China in 1944–45. However, Soviet intervention in August 1945, which is not hinted at in the map, was to make a major impact. The advance by Nimitz in the Central Pacific was strategically more significant than that by MacArthur in the South West Pacific. Moreover, there was a major contrast between the central Pacific, where the Marines faced well-prepared defences on relatively obvious targets, whereas, in the south-west Pacific, the focus was on surprise landings on relatively lightly-defended beaches across a range of possible targets. In both cases, naval superiority was crucial as the offshore fleet was immobile while protecting the landing parties. ABOVE

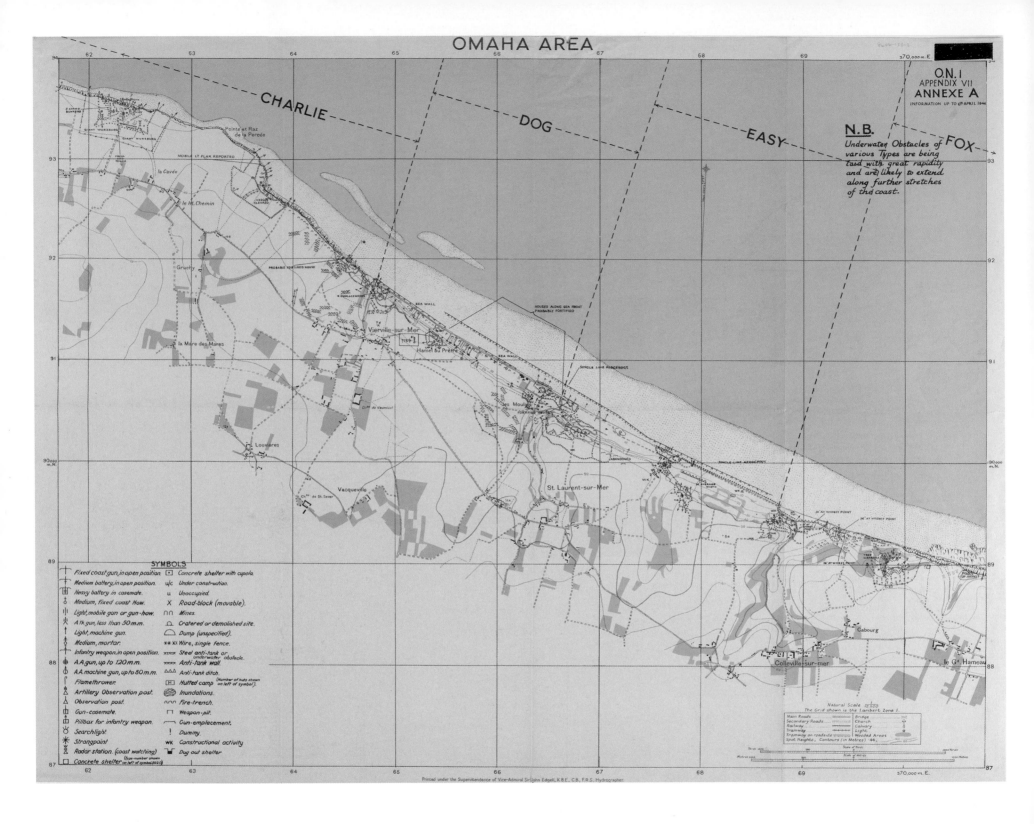

MAP OF OMAHA BEACH ILLUSTRATING THE NATURE OF GERMAN DEFENCES, 1944 Although heavy casualties were suffered by the Americans in storming the German defences, the Allied invasion force benefited from well-organised and effective naval support, as well as from absolute air superiority. The Anglo-American naval armada prevented disruption by German warships. There was no equivalent to the challenge posed by the Japanese fleet to the American landings in the Philippines later that year, but attacks by destroyers, torpedo boats and submarines based in French Atlantic ports were a threat to the landing fleet and to subsequent supply shipping that had to be guarded against, and this was done successfully. On land, the topography was a major issue, notably the bluffs behind the beach and the narrow nature of the beach area. The Americans had failed to learn sufficient lessons from their amphibious assaults in the Pacific and Allied operations in the Mediterranean. **ABOVE**

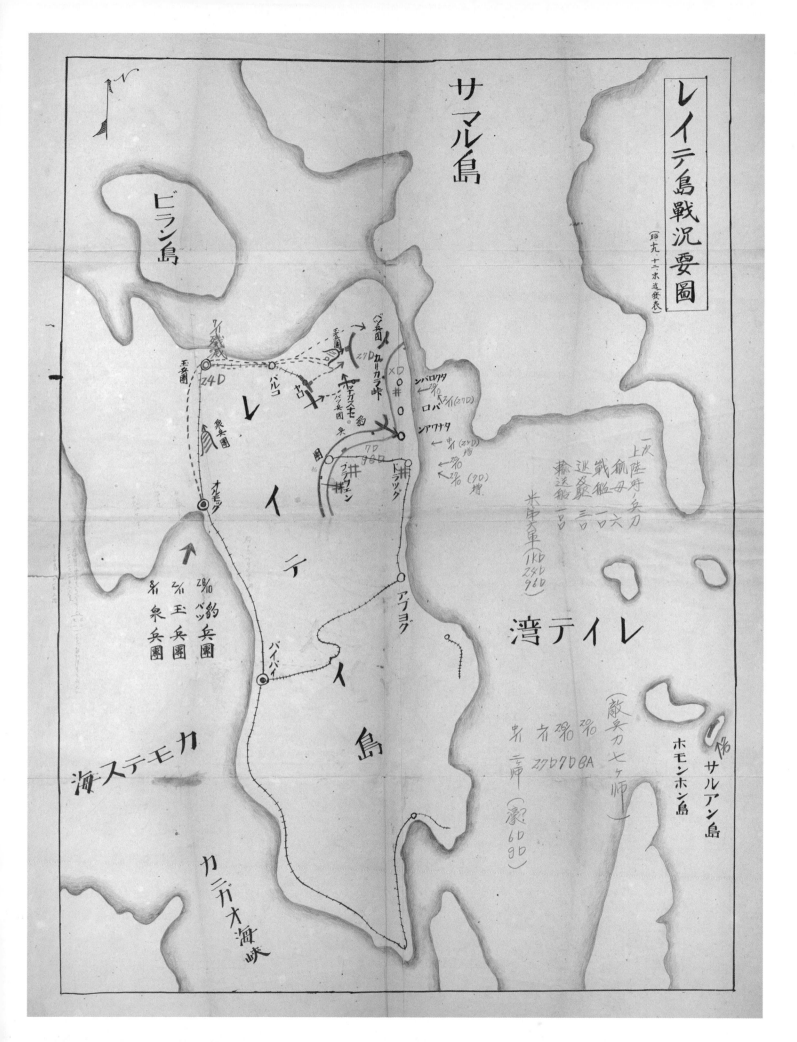

JAPANESE MAP OF THE BATTLE OF LEYTE GULF, 1944
Conflict in the Philippines in late October 1944 was the background to and consequence of the American victory over the Japanese in the Battle of Leyte Gulf. This, the largest battle of the war, arose from an overly complex Japanese scheme to attack the vulnerable American landing fleet, one that posed serious problems for the ability of American admirals to read the battle and control the tempo of the fighting. In this Japanese diagram, Japanese units are shown in blue and American ones in red while the positions of the respective fleets are indicated by the characters in the sea. American forces landed on the island of Leyte from 24 October only for the Japanese to send reinforcements. In response the Americans mounted an additional landing on the west coast of the island on 7 December. LEFT

TAIHEIYŌ SENSŌ NIHON KAIGUN KANSEN SŌSHITSU ICHIRANZU (1947) Map of the western Pacific showing Japanese naval losses during the Second World War. Includes tables and graphs of shipping data. The significance of the Solomon Islands and the Philippines emerge clearly. Losses in the more northern parts of the Pacific were limited, although the fighting over the Aleutians was responsible for a few losses. The Japanese navy was eventually totally outfought by that of the USA. The Japanese continued building warships, but their numbers were insufficient and their navy lacked the capacity to resist the effective American assault. It also suffered from poor doctrine, including a lack of understanding of naval air war and an inadequate appreciation of respective strategic options. The Japanese navy proved particularly inept at anti-submarine operations. The navy also suffered as the army became more powerful. **RIGHT**

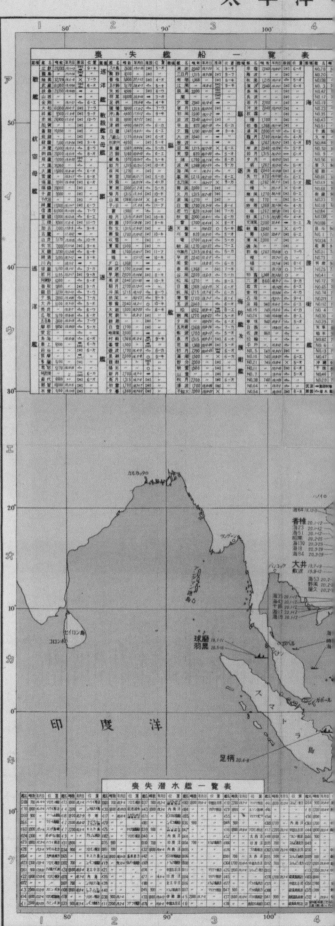

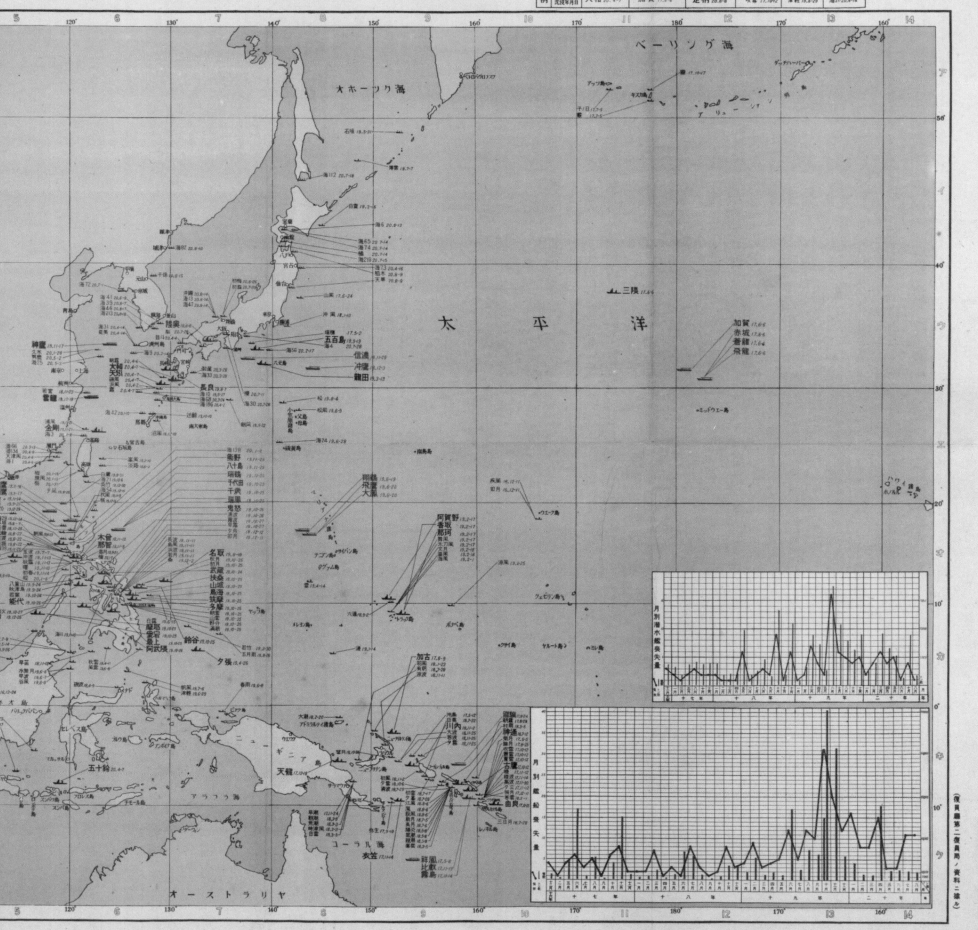

INCHON OPERATIONS, 1950 A map showing the movement of American forces from Japan and the Pusan Perimeter prior to the Inchon landing, a key episode in the Korean War. American naval dominance was a vital prerequisite for such an operation, not least due to proximity to China, which included Soviet naval bases in Manchuria. This daring and unrehearsed landing in very difficult tidal conditions applied American force at a decisive point. Carried out far behind the front, and with very limited information about the conditions, physical and military, that they would encounter, about 83,000 troops were successfully landed. They pressed on to capture nearby Seoul. The significance of bases in Japan emerges clearly. **OPPOSITE**

THE COLD WAR 1946–1989

Knowledge of the oceans was transformed during the Cold War, in part due to the military confrontation of the period between American-led and Soviet-led coalitions, and in part to the development and application on a global scale of advanced technology. Before the twentieth century, knowledge of the deep seas had been limited, although the laying of telegraph cables in the nineteenth had brought some information about the oceanic floor.

In the twentieth century, the increased scope of knowledge came in a number of ways: from aircraft, satellites, submersibles, surface ships using sonar, and from boring into the ocean floor. The ocean floor's effect on the water surface, and its contours, could be picked up on radar images taken from aircraft and satellites. Water temperatures, measured in the same way, provided warning of forthcoming storms. Ship- and air-borne towed magnetometers and deep ocean borehole core sequences gathered widespread data about magnetic anomalies. Submersibles able to resist extreme pressure took explorers to the depths of the ocean. In 1960, Jacques Piccard, accompanied by Don Walsh, an American naval lieutenant, explored the deepest part of the world's oceans, the Marianas Trench near the Philippines, in the bathyscaph *Trieste*, providing precise information about a location that had hitherto served as a source for mysterious rumours about strange creatures and other abnormalities.

From the 1970s, metals on the ocean floors were mapped. Unmanned submersibles with remote-controlled equipment furthered underwater exploration and mapping. Thanks to such information, mapping of the ocean floors became more ambitious. Thermal hotspots on ocean floors were charted and an understanding of their causes and impact developed. The role of the thermal layer on the effectiveness of

sonic systems encouraged charting. By the end of the century, a full map of the ocean floors was possible, one that revealed the value of the Seasat imagery produced from 1978.

NEW DEVELOPMENTS

Such activity needs to be placed in context. A key element was provided by the development of submarine capability in the Cold War, as the somewhat limited submersibles of the Second World War were transformed into large ocean-going submarines able to deliver formidable weapons, including sub-surface launched missiles which, eventually, carried nuclear warheads and had considerable range.

The resulting capabilities created opportunities and threats. The United States and its allies, notably the NATO powers and Japan, were keen to develop information to aid their understanding of how best to scrutinise the movements of Soviet submarines from their bases in the Kola Peninsula near Murmansk and at Vladivostok into the open oceans. Secret underwater mapping and hydrography were important to the establishment of patrol areas for submarines and to communications with them, as well as for the location and maintenance of far-flung systems of underwater listening devices.

This situation led to a major expansion in financial support for marine sciences. The NATO framework included a Science Committee. Oceanography as a subject and institutional structure grew significantly from the 1960s, although some influential oceanographers, such as George Deacon and Henry Stommel, were troubled about the consequences of military links.

Not only were the Western powers active. The Soviet Union established a Marine Cartographic

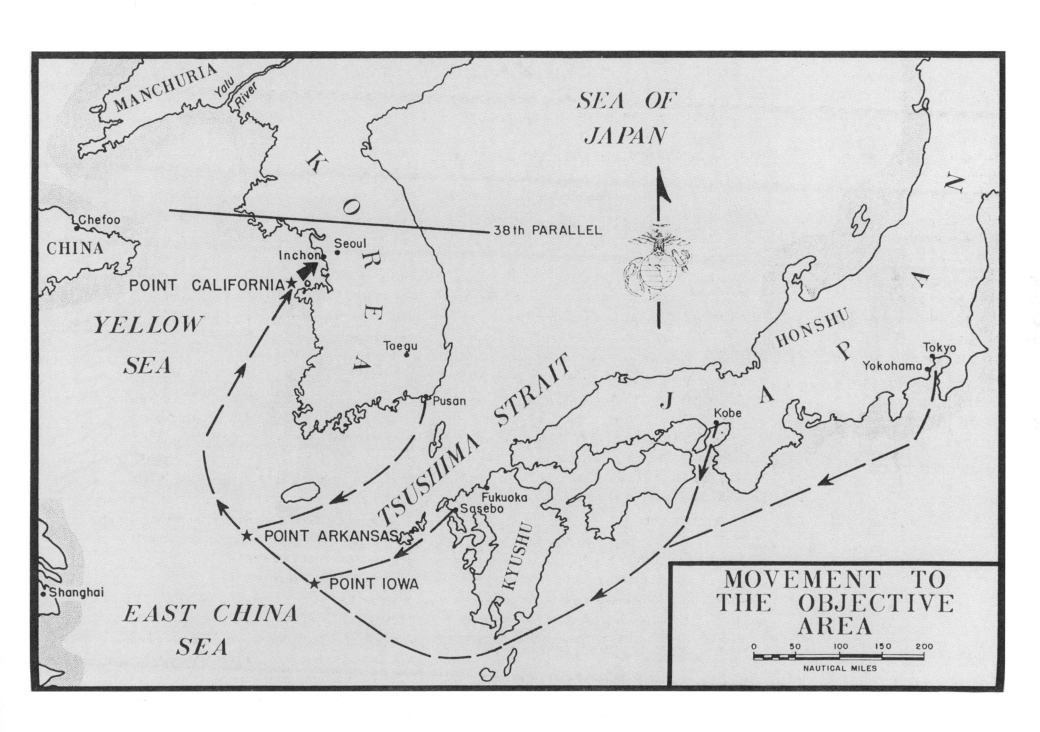

MOVEMENT TO
THE OBJECTIVE
AREA

0   50   100   150   200
NAUTICAL MILES

CUBAN MISSILE CRISIS, 1962 The disposition of American forces as of 22 October 1962, indicating the range of American naval facilities. That day, in response to information obtained by a spy plane on 1 October that medium-range nuclear missile sites were under construction in Cuba, a breach of Soviet assurances, President Kennedy addressed the American people. He announced a 'strict quarantine on all offensive military equipment under shipment to Cuba', in other words a blockade, which took effect on 25 October. The Americans prepared for air attacks on Cuba and an invasion, as well as for nuclear strikes against the Soviet Union. **RIGHT**

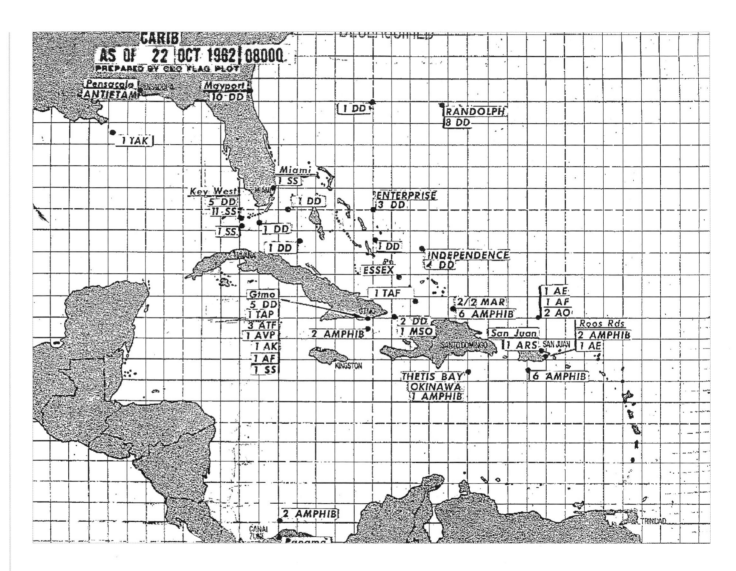

Institute in 1946 and it focused on foreign waters in the NATO area. By 1955, the northern hemisphere had been covered by Soviet navigational charts and, thereafter, the focus was on the southern hemisphere, which reflected the greater range of Soviet naval ambitions. Soviet maps, however, were not as good as their American counterparts.

Naval mapping took a number of forms and had a range of consequences. The American blockade during the Cuban Missile Crisis of 1962 was a prime example

of the role that naval maps played in the Cold War. The American 'exclusion zone' had to be declared in a way that was understandable to all involved, even those in a submerged Soviet submarine.

Naval mapping included the use of submarines, surface warships and naval aircraft for espionage purposes, to accumulate information on enemy communication systems with a view to opposing them. In 1960, the American navy began to build 'technical research ships', equipped with radio interception devices and operating as seaborne listening posts. The first, the USS *Oxford*, was used in 1961 to intercept microwave telephone communications from Cuba, which was regarded as important in understanding the direction of Soviet policy. From 1962, the American navy ran patrols along the Chinese coast in order to acquire information by radio interception. A similar mission by the USS *Maddox* along the North Vietnamese coast, designed to locate all coastal radar transmitters, led in 1964 to the Gulf of Tonkin incident, an unsuccessful attack by North Vietnamese torpedo boats that precipitated a far higher level of American engagement in the Vietnam War. In 1968, another spy ship, the USS *Pueblo*, was captured off North Korea, leading to a major incident. The Soviet Union similarly had listening posts disguised as trawlers.

THE CHANGING ROLE OF NAVAL POWER

The Cold War saw no large-scale naval clashes on the scale of the world wars, but the role of naval power was important to it. This, however, had not been apparent at the outset. The largest war since the Second World War, and the first major clash of the Cold War, the Chinese Civil War of 1946–49, was not one in which naval conflict played any significant role, and the side without a navy, the Communist side, won. Their lack of amphibious capability became more significant when offshore targets were attacked: it was partly responsible for the total failure of the Communist assault on the Quemoy archipelago in October 1949.

A similar lack of capability hampered the challenge of Communist power. Soviet ground forces were the most important in Eurasia, and threatened to overrun Western Europe. The American counter was atomic weaponry, which had been used so effectively against

Japan in 1945, but in the late 1940s and 1950s this was seen as best delivered by aircraft. As a result, the expansion plans of the American navy were dramatically scaled back in the late 1940s. The same was also the case for the Royal Navy, the world's second largest at the close of the Second World War, and for the Canadian navy, the third largest. Their expansion plans were ditched, while the size of the navies was cut.

The Korean War (1950–53) both confirmed and challenged this assessment. The decisive fighting took place on land and the American air force played a major role in support of United Nations forces. However, the navy (alongside those of its allies, notably Britain) also provided important backing, especially for carrier air intervention, which provided the early air response, and for shore bombardment. In addition, the Americans mounted a major amphibious attack, at Inchon in 1950, that helped determine the sway of campaigning for a while by threatening North Korean communication routes. The landing of major American forces at Inchon and their advance on Seoul led the North Koreans to pull back their forces that were further south in South Korea.

The Korean War encouraged rearmament in the West, notably by the USA. This rearmament led to the building of a new series of fleet carriers designed to carry nuclear bombers. The protection of these carriers became central to American naval doctrine, but at the same time the air force remained the core military component.

In the 1960s, the situation altered. The large-scale use of American carriers to provide important air support during the Vietnam War, and the relative invulnerability and proximity of carriers compared with land bases, underlined the value of naval power for Third World operations. Warships were, for

COLD WAR IN THE ATLANTIC Operational Ranges of Soviet Submarines, British assessment, 1963. As a result of the problem of access to the oceans from the Baltic and Black Seas, the Soviet navy had developed their Northern Fleet based at Murmansk and nearby Severomorsk. Submarines had played a role in the Cuban Crisis of the previous year. Khrushchev sought to deploy to Cuba four diesel-electric submarines, each equipped with a single nuclear torpedo with a yield similar to that of the bomb dropped on Hiroshima. To force Soviet submarines to the surface during the crisis, American antisubmarine forces dropped warning depth charges, which led to the risk that their captains would fire their torpedoes. During the crisis, the problems of maintaining communications with submarines proved serious. The nuclear weapons in the submarines were under the command of the captains, which could have led to a rapid escalation of the crisis. The Soviet Union initially planned to base a third of its nuclear-armed missile submarines in a permanent navy base in Cuba at Mariel and also to develop a base for a sizeable surface fleet. **RIGHT**

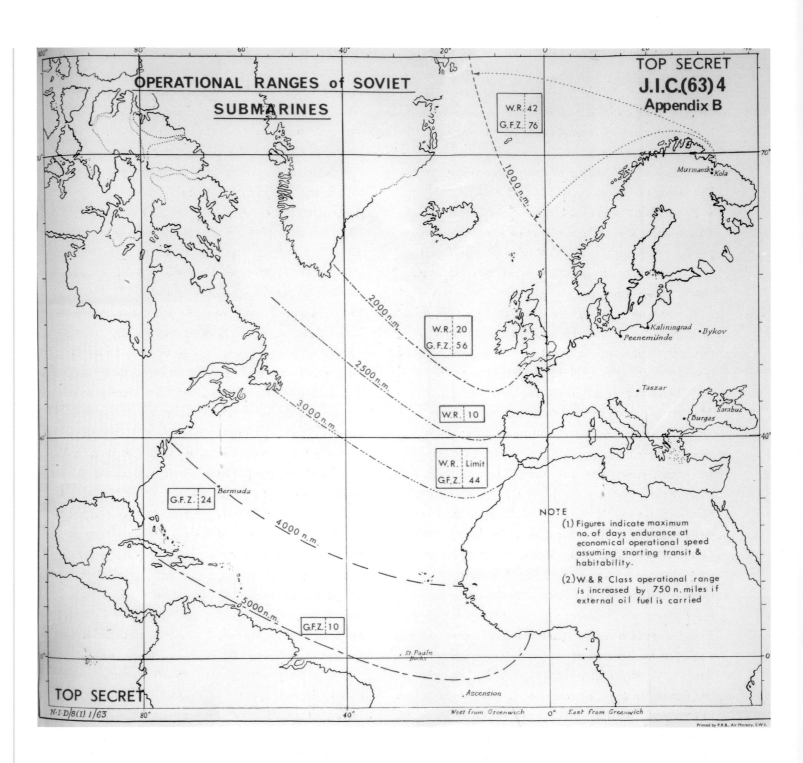

example, used to challenge the Viet Cong's employment of waterways, especially in the Mekong Delta in 1967–68. This strategy owed much to the weakness of North Korean, North Vietnamese and Chinese naval power, and to the extent to which Soviet naval power was outclassed. The deployment of American and British submarines armed with ballistic missiles provided another crucial capability, as these submarines could be sent close to the Soviet Union and China. This deployment was linked to the development of hydro-accoustic devices, such as side-scan sonars, in order to provide more accurate charts, permitting safer navigation.

At the same time, it became necessary to consider how best to confront the possibility of the Soviets doing the same with their submarines. Moreover, the development of Soviet naval power, in quantitative and qualitative terms, and its deployment in the Mediterranean from 1967 challenged American assumptions about the naval situation. There was also an asymmetric aspect: the Soviet navy put its emphasis on surface-to-surface missile ships and on submarines, and not on carriers, battleships or anti-aircraft equipment. Until the late 1970s, the Soviet fleet was designed for a battle of the first volley. In the 1962 Cuban Missile Crisis, the Soviets initially planned to deploy eight surface ships, four conventional submarines and seven missile-armed conventional submarines, but this plan was cancelled and only the four conventional submarines (each with one torpedo equipped with a nuclear warhead) were deployed. They were detected, and thus under the 'control' of the American navy. In contrast, the 1967 deployment of the Black Sea fleet into the Mediterranean seriously endangered an American carrier task force in the Mediterranean, which the Americans until then had considered their own.

COMPETITION

As a result of Soviet development, the Cold War increasingly involved naval competition. This competition was enhanced as the Soviets established and used distant bases, including in Cuba, Somalia, Syria and Vietnam. The first big visit of Soviet surface naval units to Cuba was in 1969, when one cruiser, two destroyers, a nuclear submarine, two attack submarines, a submarine tender and a tanker began what became a pattern of visits. In 1970, a submarine depot ship, a destroyer, a submarine and a tanker visited Cuba, leading to American pressure, which resulted in their departure in January 1971. From 1962, the Soviets operated research and survey vessels in the area, based at a fishery centre in Havana built in 1962–66 that had a state-of-the-art communication system and a repair facility for major 'fishing' vessels. The situation anticipated current American concern about Chinese naval plans. At this stage, China could not match this capability.

NATO sensitivities focused on the North Atlantic and the challenges posed by Soviet submarines to the ability to support NATO forces in Europe in the event of any war there. Anti-submarine warfare became a key NATO capability, including the development of hunter-killer submarines and carriers flying vertical or short take-off aircraft able to protect helicopters acting against submarines. There was, however, no war between the major powers – not since 1945 or, in the case of the leading two naval players, since 1918.

Instead, there was smaller-scale naval conflict, for example the Battle of Arafura Sea between the Dutch and the Indonesians on 15 January 1962, in which a Dutch frigate defeated an Indonesian attempt to insert troops in Dutch New Guinea, sinking a torpedo boat. This battle is honoured in Indonesia on 'Ocean Duty Day'. Naval conflicts in South and Southwest Asia were an aspect (and a secondary one at that) of conventional wars between India and Pakistan and between Israel and its Arab neighbours. The use of surface-to-surface missiles fired from missile boats was crucial, notably in 1967 when Egyptian-fired Soviet Styx missiles led to the sinking of the Israeli warship INS *Eilat*. This success, which opposed Western and Soviet equipment, underlined the extent to which American warships were not adequately prepared for the challenge of this kind of attack. Indian naval pressure on Pakistan proved an aspect of superior Indian capability.

In 1974, Chinese and South Vietnamese warships briefly clashed over the disputed Paracel Islands in the South China Sea. This was the People's Liberation Army Navy's first sea battle against a foreign foe, and a Chinese success. It was followed by comparable action against Vietnam in 1988 over the Spratly Islands. The Americans refused to provide South Vietnam with assistance in 1974, but the threat of such intervention helped restrain Chinese policy during the crisis. Both China and South Vietnam made repeated errors in 1974, including over tactics, equipment and logistics. This is more generally the case with naval conflict than tends to be appreciated. In part the situation reflects the difficulty with maintaining naval capability, which requires effective warships, weapon systems, command decisions, training and doctrine.

FALKLANDS WAR
The most significant naval warfare of this period was that between Argentina and Britain in 1982, a conflict between two Western states. The Argentinian conquest of the inadequately protected Falkland Islands, a British possession in the South Atlantic, led to a counter-invasion. This was dependent on the despatch of a naval task force and on it maintaining

operational effectiveness and gaining naval and air superiority. All this had to be achieved before the onset of the South Atlantic winter. Given the rate with which British surface naval capabilities were being discarded following the Nott Report, as the navy was being focused on NATO anti-submarine tasks and on the submarine-based nuclear deterrent, it was fortunate for Britain that the invasion was not mounted the following year, when there would have been less carrier capacity.

In 1982, the Argentinian navy was not inconsiderable, but it was prepared for combat with with the navy of neighbouring Chile and not the Royal Navy. The sinking, by the submarine HMS *Conqueror*, of the cruiser ARA *General Belgrano* (so far the largest warship sunk since the Second World War) led other Argentinian surface ships to retreat to port, including the Argentinian carrier. Thus submarines acted as a key operational deterrent although Argentinean submarines deployed against the British did not have the same impact. None, however, were sunk. Off the Falklands, British warships confronted bombs and air-launched missiles. Their lack of adequate close-in air defences and an airborne radar system to warn of over-the-horizon incoming aircraft proved major problems, causing losses, notably HMS *Sheffield*. However, this problem was eased by weaknesses in the Argentinian air assault, including British-supplied Second World War bombs that did not explode. Meanwhile, the British carriers provided the task force with air support but lacked the strength to provide air dominance. Following the sinking of the supply ship SS *Atlantic Conveyor* by Argentinian air attack, air mobility was handicapped with the loss of nearly all the Chinook heavy-lift helicopters. Although air cover remained a problem, British troops were able to land and to reconquer the islands.

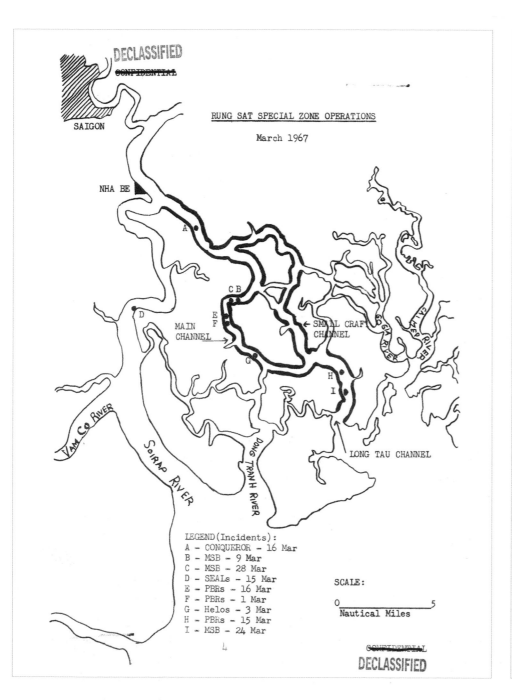

RUNG SAT SPECIAL ZONE OPERATIONS

March 1967

SAIGON

NHA BE

A

C B

E
F

D

MAIN
CHANNEL

← SMALL CRAFT
CHANNEL

G

H

I

VAM CO RIVER

SOIRAP RIVER

DONG TRANH RIVER

GO GIA RIVER

CAI MEP RIVER

LONG TAU CHANNEL

LEGEND (Incidents):
A – CONQUEROR – 16 Mar
B – MSB – 9 Mar
C – MSB – 28 Mar
D – SEALs – 15 Mar
E – PBRs – 16 Mar
F – PBRs – 1 Mar
G – Helos – 3 Mar
H – PBRs – 15 Mar
I – MSB – 24 Mar

SCALE:

0 ————— 5
Nautical Miles

4

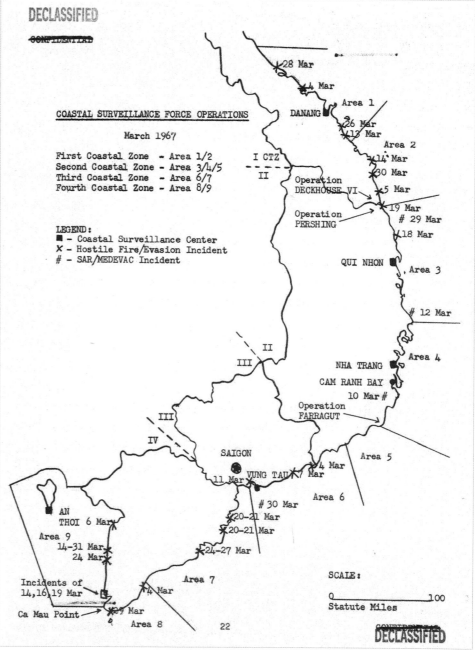

COASTAL SURVEILLANCE FORCE OPERATIONS

March 1967

First Coastal Zone – Area 1/2
Second Coastal Zone – Area 3/4/5
Third Coastal Zone – Area 6/7
Fourth Coastal Zone – Area 8/9

LEGEND:
■ – Coastal Surveillance Center
✗ – Hostile Fire/Evasion Incident
– SAR/MEDEVAC Incident

28 Mar

4 Mar

Area 1

DANANG

26 Mar
13 Mar

Area 2

14 Mar

30 Mar

I CTZ

II

Operation
DECKHOUSE VI

Operation
PERSHING

5 Mar

19 Mar

29 Mar

18 Mar

QUI NHON

Area 3

12 Mar

II

III

NHA TRANG

CAM RANH BAY

Area 4

10 Mar #

Operation
FARRAGUT

III

IV

SAIGON

11 Mar

VUNG TAU

4 Mar

Area 5

Area 6

30 Mar

AN
THOI 6 Mar

Area 9

14-31 Mar

24 Mar

20-21 Mar

20-21 Mar

24-27 Mar

Incidents of
14,16,19 Mar

4 Mar

Area 7

Ca Mau Point →

29 Mar

Area 8

22

SCALE:

0 ————— 100
Statute Miles

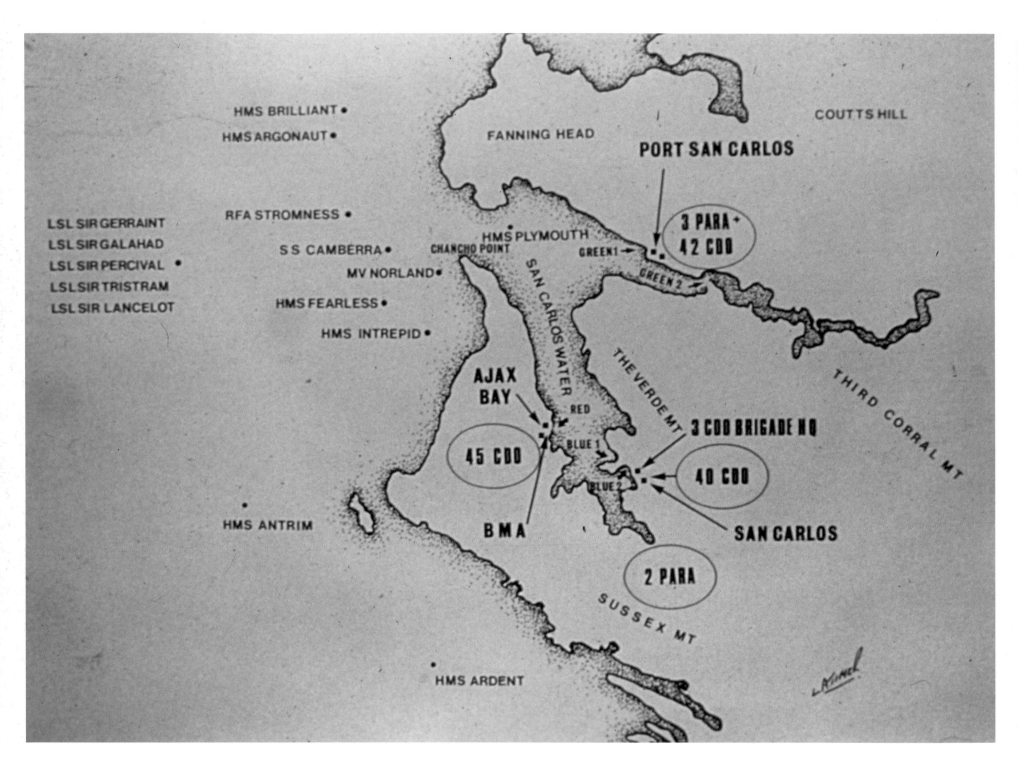

HMS BRILLIANT •

HMS ARGONAUT •

FANNING HEAD

COUTTS HILL

PORT SAN CARLOS

LSL SIR GERRAINT
LSL SIR GALAHAD
LSL SIR PERCIVAL •
LSL SIR TRISTRAM
LSL SIR LANCELOT

RFA STROMNESS •

HMS PLYMOUTH •

S S CAMBERRA •

CHANCHO POINT

MV NORLAND •

HMS FEARLESS •

HMS INTREPID •

3 PARA •
42 CDO

GREEN 1

GREEN 2

SAN CARLOS WATER

THE VERDE MT

THIRD CORRAL MT

AJAX BAY

RED

3 CDO BRIGADE HO

45 CDO

BLUE 1

40 CDO

BLUE 2

SAN CARLOS

HMS ANTRIM •

BMA

2 PARA

SUSSEX MT

HMS ARDENT •

FALKLANDS WAR, 1982 Landing zones of 3 Commando Brigade at San Carlos on 21 May 1982. The landing of British forces that day moved the conflict from being solely a sea-air war. This was necessary for the British, both to secure their objective, and because, at sea, air-launched Exocet missiles and bombs had led to the loss of a number of British ships, showing that modern antiaircraft missile systems were not necessarily a match for manned aircraft. The Royal Navy lacked, as did most units of the American navy, an effective close-weapon system against aircraft and missiles. By landing on the Falklands, the British had not ensured success, as the Argentinian plan rested on fighting on from fixed positions in order to wear down British numbers and supplies. However, British success in the field, combined with an ability to maintain the initiative, destroyed the Argentinian will to fight on. ABOVE

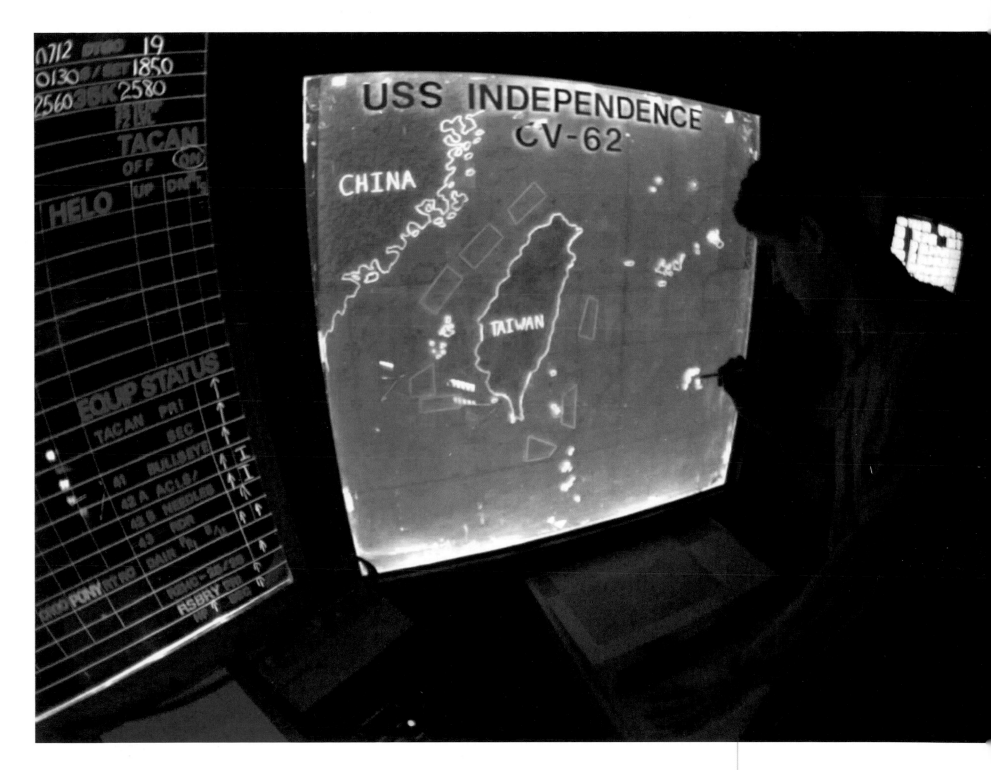

As with the successful American invasions of Grenada in 1983 and Panama in 1989, the Falklands War indicated the importance of naval forces in combined operations and therefore the enabling capability provided by mapping and charting. A commitment to combined operations played a role in the major American naval build-up of the 1980s under President Reagan, and had a later echo in

Chinese interest in developing naval capability. In the 1980s, the American navy was, however, primarily configured both to protect transatlantic maritime routes and to attack the Soviet Union, projecting force into Soviet waters. It did so in exercises, notably that of 1982, in which an American attack on the Kola Peninsula was simulated.

LOCATING POSITION, 1996 Air Controller on the USS *Independence* checks the current position of the aircraft carrier on an electronic map which keeps track both of ships and of its air wing. China had told the United States not to send its navy into the Taiwan Straits. **ABOVE**

However, the Cold War ended without any conflict
to test respective naval capabilities. The second-largest
navy in the world, that of the Soviet Union, played no
role in the crisis of Communist power that led to its
collapse, first in Eastern Europe in 1989, and then in
the Soviet Union in 1990–91.

NAVAL WARFARE SINCE 1990

The need for information became even more acute as
network-centric warfare demanded an understanding
of the location and moves of units over a greater spatial
range and in three dimensions on, above and under
the sea surface. Automated intelligence handling
systems drawing on computer power proved the
solution, and these systems linked surveillance,
targeting and firing. Automatic weapons control
systems, such as the American Aegis air-control
system, played a significant role.

The need for information had consequences for
the mapping of data, notably the range, scope, scale,

complexity and timeliness of data. In this context,
the notion of a usable picture of reality meant that
command and control decisions were more difficult
than in the case of the network-centric information
earlier offered by radio and, later, radar. Operationally
crucial, these systems, however, have limitations in
offering strategic insights. Moreover, the rise of
GPS-controlled navigation is only an advantage
if the GPS satellites are working, or at least sufficient
satellites are working. Ancient and tested skills of
navigation not based on GPS-like navigational
tools were in danger of being lost. This has now
been recognised and is leading to a reinvigoration of
terrestrial-based systems, such as eLoran. The increased
availability of a greater range of digital data has given
rise to better use of geographical information. Some of
the data are used for navigation, while other forms are
being further exploited for a wider variety of mapping
and visualisation requirements to understand better the
breadth of the battle space.

Naval warfare after the end of the Cold War has hitherto focused on the use by Western powers, notably the USA, of naval forces to support interventions. The most significant were the Gulf Wars of 1991 and 2003, but the names Afghanistan, Bosnia, Kosovo, Sierra Leone and Somalia are only the most prominent of a sequence. Others include Haiti, Liberia and, more recently in 2011, Libya. A key role in the face of opposition has been the delivery of ordnance, notably by carrier aircraft and by submarine- and surface-launched cruise missiles. The provision of amphibious capability has also been important, and has been enhanced by the extensive use of helicopters. As a consequence, naval power becomes part of the equation in inland countries, such as Afghanistan. In 2001, American helicopters flew 450 miles from the Arabian Sea to Kandahar. This capability was particularly evident in the 2003 American-led invasion of Iraq. The absence of opposing navies, or naval forces of any scale, ensured that ship-to-ship operations were limited or non-existent.

Naval strength in these respects led to an underplaying of the potential significance of large-scale naval warfare with other naval powers. This underplaying, which focused on American dominance, was encouraged by the rapid decline of the Russian (formerly Soviet) navy after the Cold War. It lost bases with the break-up of the Soviet Union and resources, not least as the Russian government faced serious fiscal problems and focused on the army, especially as a result of its counter-insurgency war in Chechnya in the Caucasus, an inland war. Much of the navy, especially the surface navy, rotted away, and radioactive leakage from poorly maintained submarines and surface units caused an environmental disaster in the Murmansk region.

Major efforts were made to keep the submarine fleet effective, but the loss of the submarine *Kursk* in an accident in 2000 underlined a general sense that the Russian navy had lost effectiveness.

NAVAL RACE
The situation changed from the 2000s as China and Russia made major attempts to build up their fleets, develop new capabilities and acquire significant asymmetrical advantages over the USA. China was determined to prevent the USA from continuing to thwart any Chinese action in the Taiwan Strait. China's wealth and regional ambitions posed particular problems for the USA, as well as for regional powers, and led to a naval race, with Japan, Australia and India, in particular, building up their navies in response to apparent Chinese plans, as did Vietnam and other regional powers. The acquisition of submarines was a key focus, although India also built up a carrier force. In order to support its allies, the USA in the 2010s

POLICING A TROUBLED FRONTIER, 2003 A South Korean naval officer explains, using a map, the navy's warning shots fired on North Korean fishing boats after they crossed the disputed maritime border. ABOVE sought to 'pivot' its forces towards the Asia–Pacific region. China treated American actions as provocative challenges to its sovereignty in seas that it claimed were Chinese.

There is notable tension in the East and South China Seas. China seeks to limit or threaten access to these waters by navies it judges hostile, such as that of the USA, and thus control maritime access to China and to other regional powers, as well as dominate one of the leading seaways of the world. The Chinese are building islands in the middle of the open sea and installing defences on them. In part, the naval effort is part of a long-standing Chinese threat to Taiwan, but it has broadened out greatly. In addition, Chinese interest in port facilities, notably in Pakistan, Sri Lanka and Djibouti, has led to concern about Chinese naval plans in the Indian Ocean and has encouraged India's development of its naval capability. Russia meanwhile has resumed sending warships to Cam Ranh Bay in Vietnam, a Soviet base during the Cold War. China derives much of its oil and raw materials from the Indian Ocean. Its economic activity in Africa has made routes through the Indian Ocean more significant. This activity has threatened the previous strategic situation in which the USA dominated the Indian Ocean while India was the key other power. Prestige as well as power projection has encouraged India, like China, to put an emphasis on developing a carrier force which, in turn, provides additional targets for submarine attacks.

Russia also has invested heavily in a new naval capability and, in particular, revived its submarine force. This poses problems for the NATO powers as they have run down their anti-submarine capabilities. However, President Putin's bold talk of launching several carriers appears less convincing due to the lack of necessary shipbuilding capability and the downturn

in the Soviet economy in the mid-2010s as oil prices fell greatly. The Syrian crisis in the 2010s had a naval dimension, with Russia seeking to protect its Mediterranean naval base in Syria and also deploying anti-ship missiles there that threatened other regional powers and that matched China's development of an anti-access, sea-denial capability.

In 2016, the Russians sent their sole carrier, the *Admiral Kuznetsov*, into the Mediterranean as part of a battle group that included Russia's largest battle cruiser, a frigate and several submarines. The carrier dispatched aircraft against Syrian rebel targets, notably in Aleppo. These were the first carrier-based combat sorties in Russian military history. The frigate launched Kalibr cruise missiles. However, the Russian carrier faced serious operational difficulties; with its engines, due to the lack of a launching catapult, and thanks to the issue of pilot skill. As a result, most of the aircraft were speedily transferred to a land base. One of its four Mig-29s crashed into the sea when it tried to land on the carrier. At the same time, the Americans had two carrier strike groups in the Mediterranean.

In contrast to their submarines, the Russians lack a strong surface fleet, which affects Russian power projection. Russian operations against Georgia in 2008, in Crimea in 2014, and against Ukraine in 2015 were all land-based. The eventual refusal, as a result of Russian policy in Crimea, of France to sell two helicopter carriers proved highly significant. Germany also refused to supply turbines to Russian frigates: the Russians have a major problem with ship turbines. Moreover, Russia lacks recent anti-submarine surface units to hunt enemy submarines and to protect its own.

Russia and China have co-operated in joint naval exercises and in a flourishing bilateral arms business, which includes the sale of Russian naval expertise to

China. They also stand to co-operate in developing the North-East Passage from the Atlantic to the Pacific north of Russia, which is expanding and easing thanks to the melting of the Arctic ice. This last change, which also affects the North-West Passage, opens up the possibility of a greatly expanded sphere of naval power politics and confrontation. It therefore ensures that the mapping of naval possibilities and prospects has to include this and other indices of climate change. The upgrading of regional capability was seen with Russia, more reluctantly, Canada and with Norway which, in 2011, took delivery of the five Nansen-class helicopter-carrying frigates decided on in 1999, providing an anti-submarine capability directed against Russia. The largest Norwegian military procurement programme, these ships were designed to protect the 200-mile territorial limits.

In the 2010s, American warships were increasingly buzzed by Russian aircraft, notably in the Baltic and Black Seas. In the Persian Gulf, Iran sought to exert similar pressure. In 2016, Iran also tried to develop a fleet presence in the Indian Ocean as well as the Chief of the General Staff, General Mohammad Bagheri, discussing the future prospects of naval bases in Yemen and Syria.

TECHNOLOGICAL CONTEXT

The possibility of naval conflict between all or some of the powers already mentioned currently dominates discussion of future naval warfare. There are also possible clashes between other powers, for example, Israel and its neighbours. The development of more sophisticated anti-ship missiles, notably by China, of drones, including those launched underwater by submarines, and of cyber-capabilities that will hit the software of opposing weapons systems, are all aspects of a dynamic technological context. The same applies

to new forms of surface firepower, such as guns using an electromagnetic effect to propel a projectile, which possibly increases the range and speed of the artillery, although without being a match for a good long-range missile. Promoted and known under the name 'Railguns', they are relatively advanced and could be deployed in the near future. An electromagnetic pulse could potentially be fired from a cruise missile, but this is not as far advanced. The Chinese talk about electromagnetic aspects in a naval war, suggesting they will deploy and/or resist such weapons.

At the same time, most naval powers, including the USA, Britain, China and France, have found that ships and related weapon systems have come in over budget and/or with compromised effectiveness. This is the

PLOTTING THE OPERATIONAL SPACE, 2013

Captain Jon Rodgers, commander of the USS *Ponce*, an Afloat Forward Staging Base, stands over a map on the command deck, somewhere in the Arabian Sea on May 14, 2013, on the second day of an international mines countermeasures exercise. Mines indeed pose an important asymmetrical threat. Having downgraded minesweeping skills in the 1990s after the end of the Cold War, the USA sought to rebuild them in the 2010s. However, missions now carried out by mine warfare ships will in future be performed instead by LCSs (Littoral Combat Ships) equipped with the LCS mine warfare mission package. This is mistaken as the hulls of LCSs are not built for mine warfare, while the skill set will be compromised by a lack of specialisation. **RIGHT**

case, for example, with the propulsion systems on new British frigates, the launch systems on the new American and British carriers and the range of the American F-35 aircraft designed for American and British carriers. In addition, it is generally believed that Chinese warships will struggle to match their projected capabilities and that Russian ship support and maintenance systems continue to be inadequate.

To depict or map such issues of quality, and their operational consequences, is far from easy. This factor undermines the validity of focusing simply on the number and listed firepower of warships, as does the variability of naval strategy, which alters with the

specific interests and assumptions, or strategic culture, of individual states.

These and other elements underline the difficulties of mapping naval warfare, irrespective of the major improvements in naval techniques and technology, as well as the great enhancement of in-time surveillance. The contrast between improvements and continuing problems captures many of the issues involved in thinking strategically and in providing operational effectiveness. The range of naval issues in dispute at present underlines the significance of these factors. The future is unclear, but the ability to use the seas will continue to be important in human history.

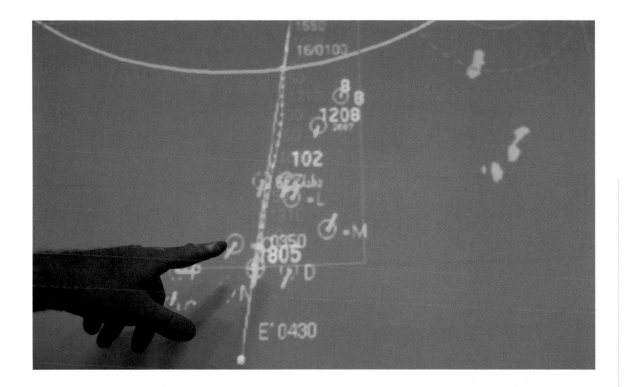

TENSION IN THE EAST CHINA SEA, 2013 A naval officer points to a navigation map showing the sailing route of a Taiwanese naval and coastguard fleet participating in manoeuvres held in the Bashi Channel on 16 May 2013. That day, Taiwan staged manoeuvres near the northern Philippines in response to the killing of a Taiwanese fisherman, after rejecting repeated apologies for the death. LEFT

CRISIS IN 2016 Markedly rising Chinese assertiveness in the East and South China Seas in the 2010s led to varied responses by other powers. Territorial claims and naval preparedness were matched by a highly precise understanding of particular interests. This photograph shows a crew of the Indonesian Ministry of Maritime Affairs and Fisheries working on a map during a patrol in the South China Sea on 17 August 2016. Indonesia sought to maintain its sovereignty over the Natuna Islands through security patrols along its Exclusive Economic Zone, but the latter overlapped with China's nine-dash line which encompasses the area China claims as its traditional fishing grounds. After a series of face-offs between government vessels and illegal fisherman, the Indonesian government sank sixty illegal fishing boats to mark Independence Day in 2016 while announcing that it sought to rename the South China Sea as the Natuna Sea in the area within 200 nautical miles of the Natuna Island, where the Indonesians developed defence facilities. LEFT

LIST OF MAPS AND ILLUSTRATIONS

INDEX

IMAGE CREDITS

2 Library of Congress, Washington DC

8–9 Fine Art Images/Heritage Images/ Getty Images

11 © British Library Board. All Rights Reserved/Bridgeman Images

12 Library of Congress, Washington DC

13 Library of Congress, Washington DC

15 DEA/A. DAGLI ORTI/Getty Images

16 DeAgostini Picture Library/Scala, Florence

19 © British Library Board. All Rights Reserved/Bridgeman Images

21 © BnF, Dist. RMN-Grand Palais / image BnF

23 © BPK, Berlin, Dist. RMN-Grand Palais / image BPK

24–25 Library of Congress, Washington DC

26–27 Stefano Bianchetti/Corbis via Getty Images

29 British Library Robana/REX/Shutterstock

30 © British Library Board. All Rights Reserved/Bridgeman Images

31 Photo Scala, Florence

32 The National Archives

33 © British Library Board. All Rights Reserved/Bridgeman Images

34 Buyenlarge/Getty Images

35 Buyenlarge/Getty Images

36–37 British Library/Robana/REX/ Shutterstock

38 Rijks Museum

39 National Library of Australia

40-41 Images@Alteamaps.com

42 Rijks Museum

43 © British Library Board. All Rights Reserved/Bridgeman Images

44 Scheepvart Museum

45 © National Maritime Museum, Greenwich, London

46 Bibliothèque et Archives nationales du Québec

47 Bibliothèque nationale de France

48-49 Fine Art Images / Heritage Images / Getty Images

51 Danish Defence Library

52 Rijks Museum

55 Bibliotheque Nationale, Paris, France / Archives Charmet / Bridgeman Images

56 akg-images / Universal Images Group / Universal History Archive

59 Universal Images Group/Universal History Archive/akg-images

60 Danish Defence Library

61 Danish Defence Library

62 Danish Defence Library

63 Images@Alteamaps.com

64 National Maritime Museum, London, UK/De Agostini Picture Library/ Bridgeman Images

66 Images@Alteamaps.com

67 Library of Congress, Washington DC

68 Library of Congress, Washington DC

69 Library of Congress, Washington DC

70 Library of Congress, Washington DC

71 akg-images / Album / Oronoz

72 Rijks Museum

73 MPI/Getty Images

74 National Library of Australia

75 National Museum of the Royal Navy, Portsmouth, Hampshire, UK / Bridgeman Images

76-77 Francis G. Mayer/Corbis/VCG via Getty Images

79 Danish Defence Library

80-81 Danish Defence Library

83 Bibliotheque Nationale, Paris, France / Bridgeman Images

85 © National Maritime Museum, Greenwich, London

87 Fine Art Images/Heritage Images/ Getty Images

88 Library of Congress, Washington DC

90 © National Maritime Museum, Greenwich, London

91 Images@Alteamaps.com

92 Images@Alteamaps.com

93 The National Archives

94 Fine Art Images/Heritage Images/ Getty Images

95 Library of Congress, Washington DC

96-97 Library of Congress, Washington DC

98 Library of Congress, Washington DC

99 Library of Congress, Washington DC

100 Library of Congress, Washington DC

101 Library of Congress, Washington DC

102 Library of Congress, Washington DC

103 Library of Congress, Washington DC

104 The Print Collector/Print Collector/ Getty Images

105 © Look and Learn/Illustrated Papers Collection/Bridgeman Images

106-107 Fine Art Images / Heritage Images / Getty Images

109 © British Library Board. All Rights Reserved/Bridgeman Images

110 Chronicle/Alamy Stock Photo

113 Private Collection / Bridgeman Images

115 National Library of Australia

117 © National Maritime Museum, Greenwich, London

118 The National Archives

119 The National Archives

120 The National Archives

122 The National Archives

123 (ART_004275_A_1) Imperial War Museum

124 (ART 4293 a) Imperial War Museum

125 www.naval-history.net

126 National Library of Australia

127 The National Archives

128 The National Archives

129 Universal History Archive/UIG via Getty Images

130 www.naval-history.net

131 Naval History and Heritage Command

132-133 MPI / Getty Images

134-135 National Library of Australia

137 The National Archives

138 The National Archives

139 Imperial War Museum

140 The National Archives

143 The National Archives

144 The Print Collector/Print Collector/ Getty Images

146-147 The National Archives

149 National Library of Australia

151 The National Archives

152 Bettmann/Getty Images

153 Naval History and Heritage Command

154 Antiqua Print Gallery / Alamy Stock Photo

155 National Library of Australia

156 David Rumsey Map Collection, www. davidrumsey.com

157 Museum of New Zealand Te Papa Tongarewa, Wellington, New Zealand/Gift of Mr. C. H. Andrews, 1967/Bridgeman Images

158 The National Archives

159 National Library of Australia

160-161 The National Archives

162-163 David Rumsey Map Collection, www.davidrumsey.com

164 Library of Congress, Washington DC

165 National Library of Australia

166 National Army Museum, London/ Bridgeman Images

167 Library of Congress, Washington DC

168-169 National Library of Australia

170-171 Purestock/Getty Images

173 Naval History and Heritage Command

174 National Security Archive/Naval History and Heritage Command

176 The National Archives

178 Naval History and Heritage Command

180 Imperial War Museum

181 Eriko Sugita/Reuters

182 Stephanie McGehee/Reuters

183 Kim Kyung Hoon/Reuters

185 Hamed Jafarnejad/AFP/Getty Images

186 Marwan Naamani/AFP/Getty Images

187a Sam Yeh/AFP/Getty Images

187b Ulet Ifansasti/Getty Images

Bridge
ASHLEY
WAPPO
FORT
CHARLESTON
COOPER RIVER
Shute
Ram
Castle
RIVER
Battery
Ram
Ft R
RIVER
Torpedoes
Ft JOHNSON
Marsh
STONO
JAMES
ISLAND
channel
MORRI
ISLAND
References
Rebel Batteries
vincents
Creek